AF522295

Global Envirnmental Governance

Global Envirnmental Governance

Tanuj Sharma

RANDOM PUBLICATIONS
NEW DELHI (INDIA)

Global Envirnmental Governance

ISBN 978-93-5111-472-7

Published in 2015 in India by

RANDOM PUBLICATIONS

4376-A/4B, Gali Murari Lal, Ansari Road
New Delhi-110 002
Phone : +9111-43580356, 011-23289044, 011-43142548
e-mail: sales@randompublications.com,
info@randompublications.com, randomexports@gmail.com

Reprint 2021

Type Setting by : Friends Media, Delhi-110089
Printed at : Replika Press Pvt. Ltd.

Preface

Since environmental issues entered the international agenda in the early 1970s, global environmental politics and policies have been developing rapidly. The global environmental governance (GEG) is the sum of organizations, policy instruments, financing mechanisms, rules, procedures and norms that regulate the processes of global environmental protection. The environmental governance system we have today reflects both the successes and failures of this development. There is great awareness of environmental threats and numerous efforts have emerged to address them globally. At the same time—and partly because of the rather spectacular growth in awareness and initiatives—the GEG system has outgrown its original design and intent. The system's high maintenance needs, its internal redundancies and its inherent inefficiencies have combined to have the perverse effect of distracting from the most important GEG goal of all—improved environmental performance.

Even though the GEG system has achieved much in the way of new treaties, more money and a more participatory and active system than anyone might have imagined three decades ago, environmental degradation continues. This book seeks to identify a number of practical steps that can foster a more efficient and effective environmental regime, making better use of the resources available and designed in a way that will be more helpful to the implementation of international environmental agreements for developing as well as developed countries. It presents up-to-date information about global environmental issues and emphasises on the importance reforming the environmental governance system. This work will be a helpful resource for students, research scholars, professors, scientists as well as for policy makers in achieving the goal of better environment.

Author

Contents

	Preface	**v**
1.	**The Changing World Environment**	**1**
	Drivers of Change	3
	The Driver-pressure Continuum	14
	Interactiion of Drivers	25
2 .	**Global Atmospheric Governance**	**31**
	International Goals and Targets	32
	Progress in Achieving Atmospheric Goals	34
	Mixed Progress	37
	Progress on Internationally agreed Goals	45
	Emerging Issues	48
	Atmospheric Governance and Integrated Approach to Management	49
	Gaps and Outlook	52
3.	**Principles of Soil Conservation and Management**	**54**
	Current State of Agricultural Land	55
	Forests	57
	Forest Plantations	58
	Drylands, Grasslands and Savannahs	59
	Wetlands	61
	Polar Regions	62
	Urban Areas and Human Infrastructure	63
	Major Issues in Land Change	64
	Competing Demands for Land	66

Separation of Consumption from the Impacts of Production 71
Land Governance 74
Outlook 77

4. Global Water Governance 82

Internationally Agreed Goals 83
Water Scarcity 84
Changes to the Hydrologic Regime Extreme Events 87
Groundwater Contamination 89
Emerging Water Quality Concerns 91
Cross-cutting Issues 92
Water Governance 98
Outlook and Gaps 103

5. Governance of Biodiversity 105

Pressures 106
Patterns of Biodiversity Change 110
Benefits to People From Biodiversity 111
Responses to the Threats to Biodiversity 115
Assessment of Progress and Gaps Conservation Strategies 122
National Biodiversity Strategies and Action Plans 123
Resource Mobilization 123
Knowledge Gaps for Biodiversity Monitoring 124
Projections, Scenarios and Horizon Scans 125
Policy Implications 125

6. Chemicals and Waste Management 127

Internationally Agreed Goals 129
Data and Indicators 131
Status and Trends of The Chemical Industry 132
Waste as an Issue of Global Significance 132
Municipal Waste 133
Life-cycle Thinking 134
Poverty and Exposure to Chemicals 135

	Marine Pollution	136
	Persistent Organic Pollutants	136
	Pesticides Including POPs	137
	Obsolete Pesticides	138
	Metals, Metalloids and Heavy Metals	139
	Radioactive Materials	141
	Emerging Issues	143
	Chemical Properties, Patterns of Use and the Environment	147
	Chemicals, Wastes and Drinking Water	148
	Reinforcing a Global Response	148
7.	**Governance of Earth System**	**150**
	The Earth System	150
	Unprecedented Changes	151
	Earth System Complexities	152
	Earth System Changes and Implications for Human Well-being	153
	Overshoot	163
	Transitions and Systemic Responses to Earth System Challenges	167
8.	**Environmental Governance in Africa**	**170**
	Policy Appraisal	172
	Transboundary Natural Resource Management	172
	Marine Managed Areas	173
	Ecosystem Services and Biodiversity Offsets	177
	Reducing Emissions from Deforestation and Forest Degradation	178
	Integrated Coastal Zone Management	179
	Sustainable Land Management	180
	Human Rights and Environmental Protection	181
	Local, Inclusive and Participatory Approaches	182
	Water Harvesting	183
	Adaptation to and Mitigation of Climate Change	184
	Stakeholder Pollution Management	186
	Building on Success	187

9. Management of Environmental Problems in Asia and Pacific **190**

Environmental Issues 190

Policy Analysis 192

Climate Change 193

Biodiversity 196

Water Resources Management 199

Chemicals and Waste 202

Environmental Governance 204

Benefits and Limitations 207

Transfer and Replication of Successful Policies 209

10. Environmental Governance Mechanisms in Europe **213**

Policy Appraisal 215

Climate Change 216

European Emissions Trading System 217

Feed-in Tarifs for Renewable Energy Systems 218

Climate Adaptation Policies 218

Air Quality 219

European Vehicle Emission and Fuel Standards 220

EU Industrial Emissions Directive 221

Local Air Quality Management Policies 222

Freshwater Management 222

Integrated Management of Transboundary River Basins 223

Policies to Address Diffuse Sources of Water Pollution 224

Water Metering and Volume-based Pricing 224

Preventions of Chemicals and Waste 226

Biodiversity Conservation 230

The Natura 2000 Network 230

Agri-environment Measures 231

11. Environmental Governance in Latin America and Carribean **235**

Regional Context 237

Effective Environmental Governance 238
Integrated Water Resources Management 241
Integrated Coastal Zone Management 245
Biodiversity Management 246
Land Use, Land Degradation and Desertification 251
Sustainable Agriculture and Livestock Production 253
Restoration of Degraded Lands 254
Climate Change 255
Reducing the Vulnerability of Populations 257
Sustainable Forest Management 259
Encouraging Diversification of the Energy Matrix 260
Enhancing Efficiency and Low-carbon Mobility 261
Policy Options and Environmental Priorities 262

12. Policy Approaches to Environmental Governance in North America 266

Market Mechanisms 267
Command-and-Control Mechanisms 271
Accountability and Transparency 273
Land use and Management 275
Integrated Watershed Management 280
Management of Energy Crisis 285

Bibliography 293

Index 297

1

The Changing World Environment

The Earth System provides the basis for all human societies and their economic activities. People need clean air to breathe, safe water to drink, healthy food to eat, energy to produce and transport goods, and natural resources that provide the raw materials for all these services. However, the 7 billion humans alive today are collectively exploiting the Earth's resources at accelerating rates and intensities that surpass the capacity of its systems to absorb wastes and neutralize the adverse effects on the environment. In fact, the depletion or degradation of several key resources has already constrained conventional development in some parts of the world.

Within the Earth System – which acts as a single, self-regulating system comprised of physical, chemical, biological and human components – the effects of human activities can be detected at a planetary scale. These have led scientists to define a new geological epoch, the Anthropocene, based on evidence that atmospheric, geological, hydrological, biological and other Earth System processes are being altered by human activity. The most readily recognized changes include a rise in global temperatures and sea levels, and ocean acidification, all associated with the increase in emissions of greenhouse gases, especially carbon dioxide and methane. Other human-induced changes include extensive deforestation and land clearance for agriculture and urbanization, causing species extinctions as natural habitats are destroyed. While humans have long been aware of the effects of their activities on the local environment, only in the last few decades has it become apparent that these activities can cumulatively affect the global environment. In the past, anthropogenic pressures on natural resources were less pervasive and the Earth's atmosphere, land and water could carry the load of human consumption and production. However, in the second half of the 20th century the effects of many diverse local changes compounded at accelerating rates to produce global consequences. Globalization allows goods to be produced under circumstances that consumers would refuse to tolerate in their own community, and permits waste to be exported out of sight, enabling people to ignore both its magnitude and its impacts. However, just as waste has – literally – reached the ends of the Earth, environmental concerns have become globalized as well.

These threats to the Earth System have led the science community and policy makers to work together more closely to meet the challenge in a sustainable and collaborative manner.

At the 1972 United Nations Conference on the Human Environment, 119 nations came together for the first time to discuss serious environmental concerns raised by the scientific and conservation communities. As an initial step, the conference established UNEP to catalyse international and UN-wide environmental action. Twenty years on, the United Nations Conference on Environment and Development in Rio de Janeiro approved Agenda 21, a blueprint for the introduction of sustainable development, a concept first articulated as "satisfying the needs of the present generation without compromising the chance for future generations to satisfy theirs" in the World Commission on Environment and Development 1987 report *Our Common Future*. In the second decade of the new century, Agenda 21 remains a vibrant and relevant guide with many of its precepts yet to be applied, particularly in regard to consumption.

The 2000 Millennium Summit, which brought world leaders together to discuss the role of the United Nations at the turn of the 21st century, produced eight Millennium Development Goals (MDGs) to make up for shortcomings that resulted from a focus on economic objectives while international development stalled. The MDGs address the integration of sustainable development principles into country policies and programmes and aim to reverse the impoverishment of human and environmental resources, while setting time-bound targets and establishing metrics. MDG 7, which specifically addresses the environment, set targets to make significant reductions in the rate of biodiversity loss by 2010, to halve the proportion of the population without sustainable access to safe drinking water and basic sanitation by 2015, and to achieve a significant improvement in the lives of at least 100 million slum dwellers by 2020.

As understanding has developed about the relationship between human well-being and environmental change, so have the attempts to make it relevant for policy makers. The dependence of social development and economic activity on environmental services and stability is increasingly understood. An economy functions within a society, or within and between societies, using natural and human resources to produce marketable goods and services. At the same time, societies survive and thrive within the environment determined by the physical limits of atmosphere, land, water, biodiversity and other material resources. Interacting environmental, social and economic forces produce a complex system that has been the focus of substantial research, but it is only in the last two decades that information and communication technologies have enabled researchers to model and explore the intricate complexities of the whole Earth System.

Insights gained from the ability to appreciate the power and nuance of Earth System complexities demand a new perception of the responsibilities and accountabilities of nation states towards planetary stewardship. This not only requires the realization of environment and development goals and targets but also the development of specific goals aimed at global sustainability, addressing the needs of the most vulnerable as well as the wants of the more powerful. The elaboration of such goals requires scientifically credible indicators and information to guide, track and report progress. Integrated environmental assessments are tools, within a broad and deep toolkit, that have been developed to meet this need. However, for the most part, policy developments and revisions have failed to adequately incorporate assessment findings and other scientific information into international policy priorities.

Drivers of Change

The last 100 years was characterized by exceptional growth both in the human population and in the size of the global economy, with the population quadrupling to 7 billion and global economic output, expressed as gross domestic product (GDP), increasing about 20-fold. This expansion has been accompanied by fundamental changes in the scale, intensity and character of society's relationship with the natural world. In tracking and analysing these transformations, a new understanding of the complexities of the Earth's biophysical systems has been developed.

It is four decades since Lovelock introduced the idea that the Earth's systems were a complex organism. More recently, science has struggled with the realization that many Earth systems are at planetary boundaries that must not be crossed. These concepts are useful to communicate both the dependence of human development on the environment and the urgency with which the consequences of collective human activity on the biological, physical and chemical processes of the Earth's systems need to be addressed. The impacts of human activities include alteration of the global carbon cycle by carbon dioxide (CO_2) and methane (CH_4) emissions; disruption of the nitrogen, phosphorous and sulphur cycles; interruptions in natural river fows that interfere with the water cycle; destruction of ecosystems that has led to the extinction of countless species; and drastic modification of the planet's land cover.

The ffifth *Global Environment Outlook (GEO-5)* is organized using the DPSIR framework consisting of drivers, pressures, states, impacts and responses along a continuum. Drivers refer to the overarching socio-economic forces that exert pressures on the state of the environment. While *GEO-4* identified drivers within a thematic context, *GEO-5* identifes two major drivers on the continuum - population and economic development - that influence cross-cutting dynamic patterns and generate complex systemic interactions. For example, the pressure of supplying food, feed and fbre to growing urban centres threatens biodiversity, a pressure then exacerbated by climate change.

Pressures can include resource extraction, land-use change and the modification and movement of organisms. For example, as economic growth and the demand for agricultural products rise, so does the conversion of land for agricultural purposes, as well as the use of agrochemicals. Similarly, market demands, trade and globalization patterns can lead to the inadvertent transport of invasive species that may wreak havoc on the natural ecosystems they newly inhabit.

The DPSIR framework asks three questions :

- What is happening to the environment and why (pressure and state)?
- What is the consequence of the changed environment (impact)?
- If appropriate, what is being done about it and how effective is it (response)?

Questions regarding the role of drivers behind pressures -and the relationship between the two - can lead to persistent theoretical discussions. *GEO-5* assumes that such roles and relationships are fuid, sometimes arbitrary, a stance that should serve the purposes of this assessment.

To effectively describe the selected drivers and for a better understanding of the pressures acting on the environment, two questions are asked that focus on why environmental changes are occurring or, more fundamentally, why there is pressure.

- What is the scale or quantity of the driver? This entails both the size of the driver and its growth rate, as well as the extent of its influence and effect on other parameters.
- What is the intensity or quality of the driver? This entails the organization of the driver as well as the various processes it exhibits and influences.

Population growth and economic development are seen as ubiquitous drivers of environmental change with particular facets exerting pressure: energy, transport, urbanization and globalization. While this list may not be exhaustive, it is useful. Understanding the growth in these drivers and the connections between them will go a long way to address their collective impact and find possible solutions, thereby preserving the environmental benefits on which human societies and economies depend.

Population

Many environmental pressures are proportional to the number of people dependent on natural resources, although technological advances can mitigate individual impacts. When a population of deer, rats or sea urchins grows beyond the carrying capacity of their ecosystem, their populations crash. Sometimes the ecosystem recovers but sometimes it is permanently altered. This has been happening to human populations for millennia as they grow beyond the capacity of their valley, island or landscape to support their society, and they face famine, plague or collapse. In the last century, as human numbers grew, people came to exploit most of Earth's surface, but it is not only the scale or quantity of the population that affes the nature of a pressure on the environment. In addition, how human populations are organized - in cities or villages, in nuclear or extended families, as migrants or those that stay behind - makes a diference to the capacity of the environment to support them in their way of life.

Quantity

The human population reached 7 billion in 2011 and is expected to reach 10 billion by 2100. Using the regions defned by the UN Statistics Division, the Asia and Oceania region has the largest population, Africa is the fastest-growing and most youthful region, and Europe and North America have the slowest-growing populations and the highest proportion of elderly. As of 2012, much of the current growth in global population can be attributed to momentum left from past population increases, shifts in generational composition, and communities with high fertility rates in rural areas of less developed countries and elsewhere. Population momentum explains the apparent contradiction between a growing population size and declining fertility rates. Higher fertility rates in previous decades have resulted in a large generation of youth now entering or in the reproductive age group. This increase in the reproducing population creates conditions for larger numbers of births overall, even though couples are having fewer children.

Fertility is declining in almost all countries, although rates vary broadly. At the global level, the crude birth rate fell from 37 births per thousand in 1950–1955 to 20 per thousand in 2005–2010, while total fertility, or the number of children per woman, declined from 4.9 in 1950–1955 to 2.6 in 2005–2010. While the fertility decline was more accentuated in developing countries – from 6.0 to 2.7 children per woman between 1950 and 2010 – fertility levels in the countries of the less developed regions are still spread over a broad range. Among developed countries, fertility levels were already relatively low in 1950 at 2.8 children per woman, but continued to fall to 1.6 children per woman in 2010, which is less than the replacement rate of 2.1 children per woman. Although the global growth rate peaked more than 40 years ago, some estimates suggest there will be another billion people by 2025 and a further billion before mid-century.

Fertility and mortality are closely linked. Fewer pregnancies, for example, translate into a reduction in maternal mortality, which in many countries is still a leading cause of death for women of childbearing age. Further, lower infant and child mortality may lead to lower fertility rates as parents become better able to depend on their children surviving.

The epidemiological transition closely mirrors the fertility aspect of the demographic transition. In regions that are in an early demographic stage – those with high birth and death rates – death clusters around infants, whose deaths are mostly related to nutritional defciencies, and those dying of communicable diseases such as influenza, malaria, tuberculosis and HIV/AIDS. In regions that have entered a later demographic stage – those with lower birth and death rates – infant mortality is low and deaths coalesce around the elderly and are associated with obesity and aging, with many deaths due to cancer and heart disease.

Mortality transitions remain distinct between developed and developing countries, despite improvements. Infant mortality has continued to decline and life expectancy to rise everywhere. Global average life expectancy in 1950–1955 was 47 years, while in 2005– 2010 it was 65–68 for men and 70 for women. There are, of course, important regional variations, particularly in terms of infant mortality in the least developed countries, young adult mortality in countries affeed by the HIV epidemic, and old-age mortality in developed countries.

Internal migration is increasingly dominated by rural-urban fows, a trend that is expected to continue. However, in some developing countries, a minority of rural-rural migrants has a disproportionate impact on tropical deforestation. Increasing migration to coastal areas and small islands can affect the environmental integrity of coastal wetlands and associated fisheries.

World population is unevenly distributed, with densities in 2010 varying from 21 000 people per km^2 in Macao to 0.03 per km^2 in Greenland. This is due to a number of factors including settlement history, regional variations in demographic dynamics such as fertility, mortality and migration, and the fact that some locations are simply less suitable for human occupation. Population is particularly concentrated at lower elevations and near coasts. An estimate from 1998 suggested that a zone below an altitude of 100 metres, comprising 15 per cent of all inhabited land, houses about 30 per cent of the human population. Low-elevation coastal zones are even more concentrated, representing about 2 per cent of total land area but housing 13 per cent of the population, and growing rapidly.

In 1950 only 29 per cent of the world population lived in urban settings and only New York and Tokyo, with their populations of more than 10 million people, qualified as megacities. The urban proportion reached 50 per cent in 2010 with 20 megacities, with the bulk of the urban population in Asia and Latin America. Urban growth rates are high in both Asia and Africa, with the highest rates in recent decades in middle-sized cities.

Quality

Beyond the size and growth rates of populations, the way people settle and the way they consume can result in effes on different ecosystems.

While all of the world's net population growth by 2050 is projected to occur in the world's poorest cities, virtually all land-cover change will take place in rural environments. The greatest human imprint on the Earth's surface has been the conversion of forest to agriculture. Currently, 37.4 per cent of the planet's land surface is used for agricultural production.

Located on only 0.5 per cent of the global terrestrial surface, urban areas' demand for food is disproportionately large in terms of world land use. At the same time, forest loss is no longer correlated to rural population growth; rather at the national scale, it is linked to the international demand for agricultural products and timber harvesting for urban consumption.

The world is nearly evenly divided between rural and urban inhabitants. One half includes rural food producers with a direct impact on land in space and time. Their effect on forests is particularly acute and widespread following rural-rural migration and the associated conversion of forests to agricultural land. This very small minority of all migrants is responsible for a significant proportion of tropical deforestation yet remains very little researched. From a drivers perspective, it is also much more difficult to manage this phenomenon due to the scale and diffuse nature of the activity. The second type is the burgeoning urban population who are concentrated in space but whose impacts on the land are indirect albeit significant.

A rising human population has also been identified as the principal root cause of the water crisis. Overall, humans use more than a quarter of terrestrial evapotranspiration for growing crops and more than half of accessible water run-of. While climate change is making some places wetter, much of Africa and the Middle East currently sufer a water scarcity that is worsening with the expanding populations. Population growth has also been implicated in water scarcity in rapidly developing countries such as China, where urban growth has exacerbated a decline in the availability of clean water by overwhelming the water supply and sanitation infrastructure.

Population is not the only problem: groundwater use is highly inequitable, for example in India where 10 per cent of large farms consume 90 per cent of groundwater. Nor is a thirsty populace the only outcome. In the Republic of Tanzania, a diverse series of drivers, including population growth, has led to water conficts. Water scarcity can also provoke migration, as documented throughout Africa.

Addressing population as a driver of global environmental change, households can be considered as units for analysing consumption patterns. In the developed world, household size is shrinking as their composition changes from extended families to nuclear ones. As a consequence, the rise in the number of households has been faster than population growth.

Research suggests that this can cause double the rise in energy consumption that would occur from population growth alone, as there is an increase in the number of appliances and the level of electricity consumed per person. Larger households generally use less energy per person than small ones, conforming to the expectations of economies of scale.

The age composition of a household also has an impact on energy consumption, Lenzen *et al.,* working with data from Australia, Brazil, Denmark, India and Japan, found that the residents' average age is positively related with per-person energy consumption, while household size and urban location are negatively associated. Transport, too, is likely to be more sensitive to the number of households, since an increase in the number of homes occurs primarily in low-density suburban landscapes, resulting in more passenger vehicles and more commuting, which add to petrol consumption and pollution.

Beyond the household unit, studies also identify impacts associated with absolute population size. A study of Californian counties found that population size significantly contributes to increases in nitrogen oxide and carbon monoxide emissions. Similarly, researchers have observed a positive relationship between population size and CO_2 emissions, with an inverted U-shaped curve relation for sulphur dioxide.

Economic Development

Consumption and production are both components of economic development and, like population, have a multiplier effect on environmental pressures. While consumption and production are technically separate socio-economic drivers, they are so inextricably linked that it is difficult to discuss them independently: the consumption of raw materials by the primary industries of mining and forestry leads to the manufacture of products that are in turn consumed by individual customers.

Quantity

The production of goods for consumption requires materials – minerals, water, food, fbre – and energy. During the 20th century, global economic output grew more than 20-fold, while materials extraction grew to almost 60 billion tonnes per year. This level of materials consumed by the human population is of the same scale as major global material fows in ecosystems, such as the amount of biomass produced annually by green plants.

Consumption and production trends appear to have stabilized in developed countries, while in emerging economies such as Brazil, China, India, and Mexico, per-person resource use and associated environmental impacts have increased since 2000, and the less developed countries are just beginning the transition towards higher consumption levels. Should global economic development continue in a business-as-usual mode and population projections persist through 2050, another sharp rise in the level of global resource use is likely.

Over the period 1970–2010, average global growth rates in GDP per person measured in purchasing power parity (ppp) fuctuated between -2 and 5 per cent annually; the average was about 3.1 per cent. Since 2001, however, China has grown at 10 per cent per year, a seven-year doubling time, and India at 8 per cent per year, a nine-year doubling time, with environmental pressures increasing at much the same pace. As a result, China is now the world's largest emitter of whether these growth rates are realistic when put in the context of the Earth System's biophysical boundaries remains to be seen.

Quality

Technology is a key factor in the production of goods and services and an important one in terms of environmental impact. It has been argued that over time, factors of intensity or quality, affeed by technological innovation, may more than compensate for the adverse effes of the rise in population, so that economic growth eventually leads to environmental improvements. An example of this is greenhouse gas emission rates in developed countries since 1970, where, it is claimed, emissions increased more slowly than economic activity because of shifts towards technologies that have a lower environmental impact. However, it is not certain whether other sectors were so successful – efforts to reduce deforestation at the national level might have shown domestic improvement, but demand may have driven increased deforestation in other countries.

The environmental Kuznets curve suggested that as countries become more afuent, concern about the environment increases, leading to policies that protect it. At the same time, preferences shift away from the most environmentally damaging goods and services.

This theory has been extensively examined and while debate continues, there seems to be clear evidence that some companies and industrial sectors have reduced their environmental impact, as the theory predicts. However, there are many obstacles to a shift towards more environmentally benign technologies: in some cases, these are economic challenges as environmentally sound technologies often have higher overall costs. But in many cases, simple cost/benefit calculations are not sufcient to explain the slow pace of growth in new technologies. For example, although researchers have noted the energy efciency gap for years, whereby economically benefcial investments in energy efciency have not been made, neither consumers nor industry have made significant investments in closing that gap despite the potentially favourable returns in energy costs saved, particularly when life-cycle costing is applied.

On the other hand, technological change that improves resource efciency can have a perverse environmental effect by decreasing the costs of resource use and thus increasing demand. If the increased demand is greater than the efciency gains, the overall consumption of a resource can actually increase, with concomitant increases in environmental impact. This phenomenon is known as the Jevons paradox or the rebound effe. The choice of technology, which is shaped by economic factors and individual and public decisions, is critical in determining the overall human impact on the environment. Research to explain the obstacles to adopting more environmentally benign, cost-effective technology is just beginning. One key factor, at least for households, is unfamiliarity with life-cycle costing and a lack of understanding of the energy and cost impacts of commonly used technologies, and it appears that the same factors may also affect organizational decision making.

Values

It is commonplace to identify values as a key driver of environmental change. At one level, the argument is straightforward: human decisions, especially about consumption, are influenced by values and those decisions have impacts on the environment. However, research on human decision making notes that values are only one element in the cognitive processes, with beliefs and norms also of great importance. While some decisions reffe a formal weighing of values and beliefs, many are made without much reffeion, on the basis of normative expectations, emotions and interpretations of symbols or quick judgements.

There is a voluminous canon of literature exploring the social psychology of environmental decision making, in which several generalizations can be discerned. First, no single factor is sufcient to explain such decisions. Values, beliefs and norms, and trust in others who must also take action or who are providing information, all matter. Second, decisions are often context-specific in the sense that individuals read the context, such as whether to emphasize a gain or a loss, and frame the decision based on that reading. Sometimes individuals act as consumers, sometimes as members of a community, sometimes as citizens. Third, social networks are of immense importance in providing context as well as shaping values, beliefs, norms, trust and other significant factors. Fourth, values, beliefs, norms, trust and other individual characteristics interact with the character of the action to be taken in shaping behaviour – for example, social psychological factors may matter little when a pro-environmental action is exceptionally easy or hard to undertake, but may be critical for actions of intermediate difficlty.

Social psychology has developed many concepts to explain the factors underlying environmental decision making. Among these, values have been explored the most thoroughly and tested empirically across many national contexts. In particular, altruism towards other humans, other species and the biosphere has consistently been found to predict pro-environmental attitudes and behaviour. In addition, a willingness to cooperate with others in experimental games, conducted in both laboratory and feld settings, varies considerably across individuals and cultures. Recently, the propensity to cooperate has been shown to matter in managing forest commons, with a substantial amount of literature showing the importance of trust in commons dilemmas. However, research on trust has not yet been linked to the larger literature on values. Consumer surveys have revealed a range of reasons why an individual is unwilling to pay more for an environmentally sensitive product. The top three reasons involve poor understanding of, or apathy towards, the negative environmental impacts of consumption decisions, while the fourth most common was whether the individual viewed an action as common practice among their peers. This last point reveals the importance of societal pressure on values and by extension how decisions that impact the environment are influenced by it.

Diets

With economic growth comes a change in dietary intensity, which Popkin describes as the nutrition transition. This happens in three states: decreased occurrence of famine with rising incomes; the emergence of chronic diet-related diseases due to changes in activity and food consumption patterns; and a stage of behavioural change where diet and activity levels are better managed for prolonged healthier lives.

The growth in food consumption and related requirements for animal feed largely determine the pace at which supplies need to grow to keep up with the domestic and export demand for agricultural goods. Urbanization, demographic change and household wealth in a number of fast-evolving regions – Brazil, China, India and Indonesia – suggest that changes in food consumption patterns are likely to have profound effes on regional food systems. These changes in consumption and consumption preferences introduce increased pressures on food and energy systems from the demand side, which forces compensating adjustments to take place on the supply side through market-mediated, price-driven interactions with producers.

As regional economies continue to grow, so, too, does the consumption and production of of meat. Livestock production is the largest anthropogenic land use, accounting for 30 per cent of the land surface of the globe and 70 per cent of all agricultural land; 33 per cent of total arable land is used for producing animal feed. Pelletier and Tyedmers suggest that, by 2050, the livestock sector alone may occupy the majority of, or significantly overshoot, recent estimates of humanity's biophysical limits within three environmental areas: climate change, reactive nitrogen mobilization, and appropriation of plant biomass at planetary scales.

As urban areas are generally wealthier than rural ones, there are considerable diferences in dietary composition, with urban diets characterized by higher levels of meat, dairy and vegetable oil. These foods are often imported and require more energy-intensive production. Globalization and urbanization are cited as causing dietary *convergence* and *adaptation*. The former refers to the focusing of caloric intake on a smaller number of staple crops, such as wheat, rice and maize, with concomitant health impacts. Dietary adaptation is characterized by a greater reliance on processed foods due to lifestyle changes, greater exposure to advertising and time constraints on food preparation. This concentration of consumption also favours the concentration of the food supply chain among a relatively small number of corporations, with an implicit preference for supermarkets and larger-scale agricultural production.

Energy-water Nexus

Another important dynamic of consumption is the trade-of between energy and water consumption. This dynamic is important for both energy production and agriculture. Gerbens-Leenes *et al.* estimate that 60–80 per cent of water used globally is dedicated to irrigation, rising to nearly 90 per cent in some low-rainfall areas. In addition, energy use for irrigation can be significant. In India, where the government often heavily subsidizes water pumping, 15–20 per cent of electricity is used for this purpose. Energy use for agriculture is considerable in both developed and developing countries, although in developed countries the energy used for processing and transporting food can be twice that of the entire agricultural production sector.

Water can also be an important resource for energy production and mineral extraction. However, freshwater pollution is a common side effect of mining, including recent hydraulic fracturing activities. China sufers from water scarcity due to a dwindling supply as well as to industrial pollution; the World Bank estimates that up to a third of water scarcity in China is due to pollution, the cost of which is equivalent to 1–3 per cent of GDP.

The Driver-pressure Continuum

As population and economic development have continued to grow despite depressions and downturns, technological innovations have enhanced the integration of communities and societies into a global civilization. Technological advances in energy and transport continually generate new opportunities for growth in production and consumption, while ingenuity applied to communication and mobility has created new goods and services that previous generations could not have imagined. The growth and integration of human settlements, societies and relationships is evidenced by rapid urbanization and globalization.

Energy

As the world population increases, more people aspire to higher material living standards – creating an ever greater demand for goods and services as well as for the energy required to provide these. From 1992 until 2008, per-person energy consumption increased at a rate of 5 per cent annually. In 2009 total global energy use decreased for the first time in 30 years – by 2.2 per cent – as a result of the financial and economic crisis; half of this occurred in the OECD countries. Oil, natural gas and nuclear power consumption all decreased while hydroelectric and renewable energy consumption increased. Coal was the only energy source that was not affeed. Primary energy consumption in 2010 is estimated to have risen by 4.7 per cent worldwide, easily surpassing the minor reduction in 2009. The rate of growth in the future, however, is expected to decrease due to an assumed levelling of population growth and continued improvements in energy efciency.

The shares of energy inputs are likely to change, with the proportion produced from oil decreasing and natural gas increasing. Coal levels are expected to stay relatively constant and nuclear energy use will increase due to investments in Asia. However, with potential policy changes following the Fukushima disaster in 2011, it is difficult to predict the growth trajectory of nuclear power. If nuclear energy plans are not followed through, more coal is likely to be used, with significant implications for climate change mitigation efforts. Developing regions show a particularly strong increase in per-person energy consumption between 2005 and 2010, although, as of 2010, this seems to be levelling of. The three major economic sectors in terms of energy consumption are:

- manufacturing: 33 percent;
- households: 29 per cent;
- transport: 26 percent.

Electricity and heat generation account for more than 40 per cent of all CO_2 emissions. Between 1992 and 2008, the annual rise in CO_2 emissions of more than 3 per cent and the total rise of 66 per cent - a much greater increase than that of the global population - was primarily the result of growth in industrial production, as well as higher living standards in many developing countries.

On a per-person basis, the largest growth in electricity production occurred in the developed countries, increasing from 8.3 megawatt hours (MWh) in 1992 to nearly 10 MWh in 2008, a diference of 1.7 MWh per person, though in percentage terms this was the smallest rise at 22 per cent. The global average per-person electricity production grew by 33 per cent, from 2.2 MWh in 1992 to 3.0 MWh in 2008, while that of developing countries grew by 68 per cent, from 1 MWh to 1.7 MWh.

In 2010, 1.44 billion people globally - around 20 per cent of the world population - were still sufering from energy poverty, without access to reliable electricity or the power grid, and entirely dependent on biomass for cooking and lighting.

The energy commodity that dominates trade volume and value is crude oil, with China continuing to rival the United States in terms of consumption. The Middle East accounts for about half of all global oil trade. Coal production increased by 3-5 per cent per year during 2005-2009, with China experiencing a 16 per cent increase in production during 2008-2009 and reaching 44 per cent of the world's total coal production of 3.05 billion tonnes. With rapidly increasing energy demand, however, China became a net importer of coal for the first time in 2007. The United States is the second largest producer of coal at 975 million tonnes per year, followed by India producing 566 million tonnes.

Renewable energy production is gaining much attention: the amount of energy produced from renewable sources, including sun, wind, water and wood, amounted to 13 per cent of the world supply in 2008, and estimates suggest 16 per cent in 2010. However, the largest renewable source is biomass at 10 per cent, with nearly two-thirds of that used in cooking and heating in developing countries. Thus, when biomass is excluded, other renewable sources provide only about 3 per cent of world energy.

There has been a 30 000 per cent rise in solar energy supply since 1992, a 6 000 per cent increase in wind energy and a 3 500 per cent rise in biofuel production, all from very low bases. This is mainly due to the decreasing cost of these technologies and the 2010 adoption by 199 countries of policies to promote renewable energy.

There has been a rapid rise in the production of biomass-based fuels for transport – from maize, sugar cane, oil palm and rapeseed. While ethanol has been widely used in Brazil for two decades, its use accelerated globally at the end of the 1990s, increasing by 20 per cent each year to reach 30 million tonnes of oil equivalent in 2009. In the early years of the 21st century, biodiesel became available, with production growing at around 60 per cent per year, reaching nearly 13 million tonnes of oil equivalent in 2009. However, recent information on biofuel production raises concerns about the direct environmental and social impacts of land clearance and conversion, the introduction of potentially invasive species, the overuse of water and the consequences for the global food market. An additional cause for concern is the purchase or leasing of land by wealthier nations to produce food and biofuels – typically in developing and sometimes semi-arid countries. This trend may have serious impacts on fossil and renewable water resources, as well as on local food security.

Investment in greening the energy sector is setting new records, totalling US$211 billion in 2010, up 32 per cent from 2009, and nearly five and a half times the 2004 figure. For the first time, new investment in utility-scale renewable energy projects in developing countries surpassed that of developed economies.

The number of nuclear power plants, seen by some as an opportunity to meet the growing demand for energy, has increased by more than 20 per cent since 1992, rising to 435 by mid-2012. According to the International Atomic Energy Agency, in the 30 countries that have nuclear power, the share of electricity generated ranges from 78 per cent in France to 2 per cent in China, which has 14 operational plants, 25 under construction and more planned. Since 1992, energy production from nuclear power sources has grown by almost 30 per cent, although the share of nuclear power in the total supply has fallen from 17.5 per cent in 1992 to 13.5 per cent in 2008. Today, around the world, 60 plants are under construction, 155 planned and 339 proposed.

Global energy consumption is expected to continue to grow. Though China's energy intensity decreased by 66 per cent between 1980 and 2002, India's energy use per unit of GDP remained relatively constant over the same period and, due to its growing economy, the country is expected to contribute 8 per cent of the world's projected growth in emissions by 2030. If the international community continues to have difficulty in addressing climate change in the near future, temperatures could increase by 3.5–6°C by the end of the century. To stem the rise in global GHG emissions, the Kyoto Protocol encouraged the transfer of cleaner technologies from developed to developing economies. Trade was assumed to be the means of distributing these technologies, but without a significant reduction in existing trade barriers, this route will have limited impact.

Serious inequities remain in meeting global demand for access to energy. Today, 1.3 billion people are lacking electricity and 2.7 billion people still rely on the traditional use of biomass for food preparation, with concomitant impacts on deforestation rates, soil erosion and human health. The reliance on fuelwood also has a demographic aspect, as per-person fuelwood consumption is shown to increase with decreasing household size but to decrease with urbanization, indicating a wealth effe. In order to achieve universal access to primary energy by 2030, an annual investment of US$48 billion is needed.

Transport

Transport serves people, production and consumption and is an important facilitator of trade. The global economy is currently recovering from a severe recession, with global industrial production and trade climbing back to pre-crisis levels, albeit with marked geographic diferences: GDP is growing fastest in China, by 10.3 per cent per year, and India, by 9.7 per cent, in 2010. Data published by Global Insight suggest that in the next 40 years Brazil, Russia, India and China (the BRIC countries) will start to approach the United States in terms of GDP, surpassing Germany, the United Kingdom, France and Italy, with the distinct possibility that China will have the world's highest GDP by 2050. This unequal growth has implications for world trade and the fow of goods, posing considerable challenges and opportunities in terms of logistics and supply chains.

Countries and entire regions appear to be specializing in their attempts to become competitive, creating even greater demand for transport. For instance, Europe, the United States, Canada and Japan are dependent on fruit exports from Central and South America, some Western European countries, many Eastern European countries, and portions of Africa. Similar differential production-consumption trends happen with all products, pushing the demand for transport even higher and making freight inelastic to fuel prices. An evolving trend to manage this ever increasing world trade is containerization, which by many in the industry is considered a major revolution in handling goods, using larger ships to achieve economies of scale. It is estimated that from 80 to 90 per cent of world trade is by sea.

In the United States, the Bureau of Transportation Statistics reports that container trade in 2005 and 2006 was double that of the previous decade, increasing to 46.3 million 20-foot-equivalent units (TEUs, 19–43 cubic metres). At the global scale, container trade tripled during the same period. The European Union (EU), the world's largest trading bloc, carries out 90 per cent of its external trade and 40 per cent of its internal trade by sea, totalling 3.5 billion tonnes. However, studies in major ports show that any environmental benefits of seafaring cargo require significant attention at the place of loading and unloading. The Port of Los Angeles in California, a major hub, has, for example, implemented a variety of policies including the introduction of cleaner trucks with refuelling stations for natural gas, performance standards for cargo handlers and harbour craft, modernized and cleaner rail locomotives, and reduced vessel speeds.

After a slump in 2008 and 2009, air freight began to return to its pre-economic-crisis levels, with annual international growth of 21 per cent in 2010, although 2011 growth is expected to be heavily dependent on consumer spending. Data from the International Transport Forum (ITF) show some recovery for rail freight but it is still sufering from the economic crisis with unknown implications for the long term; exceptionally, India continues to increase its rail freight. Similarly, recovery of road freight is very slow at the national and international levels for many OECD and ITF countries.

For passenger travel China, India and Brazil recorded 7.1 per cent growth in 2010 relative to 2009. According to the International Air Transport Association, there were 2.4 billion domestic and international passengers in 2010, approximately 6.4 per cent more than ever before, with a similar trend observed in passenger kilometres travelled. Rail passenger travel continued to decline, providing space for possible substitution by freight. Data on passenger kilometres travelled in private cars sufers poor harmonization, yet it is clear that the economic crisis reduced overall travel. Moreover, possible saturation of passenger travel by car is observed in developed economies that exhibit non-significant increases in passenger kilometres, hovering at around one digit percentage growth per year.

While transport enables human interactions that contribute to development, the infrastructure for fast, motorized means of travel also creates displacement and barriers that can divide communities and reduce well-being. Roads and the enormous amount of parking to store the world's

1 billion cars are the commonest barriers, but airports and seaports for container ships are also significant.

In societies with extremely high levels of mobility, inequities in the social distribution of related environmental pressures and benefits are of increasing concern. Because most human settlements are located close to supplies of water and agricultural land, transport infrastructure displaces food production while also fragmenting landscapes that are then less able to support wildlife. Transport also has secondary environmental impacts through expanded human access to land, as the infrastructure promotes economic activities such as mining, forestry or power generation in new locations. In addition, transport enables more extensive permanent human settlement, particularly suburban and urban growth.

Most energy for transport comes from fossil fuels, and the rise of the car has produced various specifc environmental impacts, from urban health problems through land and water degradation to contributing to climate change. Many people are optimistic about the long-term prospects for shifting to cars powered by fuel cells and electric motors, but a near-term change will be difficult, and the car is noticeably more intensive in its environmental impacts than its competitor technologies, exhibiting the highest levels of energy consumption and greenhouse gas emissions. Private car ownership can also impact patterns of urbanization by permitting dispersed and low-density sprawl, which in many contexts reffes individual household dissatisfaction with urban environments, but collectively degrades environmental quality. Like the transport infrastructure that makes them possible, these new or expanded built areas impinge on natural landscapes and amplify the direct environmental impacts of transport.

There may have been a temporary decrease in transport activity in, for example, the United Kingdom and United States due to the economic recession. However, these declines are likely to be outweighed by increases in private vehicle ownership in rapidly developing low- and middle-income nations. At present, the number of motor vehicles in the world is growing much faster than the number of people. While it is unlikely that the levels of hypermobility reached in the United States will ever be reached in many other nations, there is still massive potential for growth in the level of travel and shifts towards individual motorized vehicles, especially as incomes increase. In developing nations including China and India, the ownership and use of highly polluting motorcycles is increasing faster than cars. Even when more fuel-efcient vehicles are introduced, rising numbers may outweigh efciency benefits.

However, with aggressive moves by governments and advocacy groups in the creation of green markets, two related phenomena could emerge. The first is an offset trade market in which companies can buy offsets, as futures and options, to counterbalance their inability to manage and decrease CO_2 production. The second is an attempt to develop carbon-neutral supply chains in which the amount of CO_2 produced is offset by a variety of mitigating actions that include partnerships with the local supply chains. From a policy perspective, these could deliver some development benefit by encouraging small local producers to partner with multinational companies, helping reach carbon neutrality. Similarly, new markets are developing around a lifestyle based on promoting health, the environment, social justice, personal justice and sustainable living. Such developments ofer new policy opportunities for more sustainable development worldwide that incorporates green transport policies across all sectors.

Urbanization

Urbanization exhibits complex interactions with food, discussed earlier, and energy. Urban areas, which house half the world's population, utilize two-thirds of global energy and produce 70 per cent of global carbon emissions. The amount of energy an urban area consumes is largely dependent on the built environment – whether residential and commercial buildings or transport infrastructure. Beiing and Shanghai's rapid economic growth, for example, has been accompanied by a decrease in the proportion of emissions due to industrial activities since 1985. With the increase in personal vehicle ownership, however, emissions from transport have increased significantly, sevenfold for Beiing and eightfold for Shanghai between 1985 and 2006. This increase may, in part, have been offset by an energy-efciency labelling programme implemented by the Chinese government, credited with avoiding 1.4 billion tonnes of CO_2 emissions for 2006–2010.

In general, urban populations in developing countries generate higher greenhouse gas emissions per person than surrounding rural populations, while the opposite is true for developed countries. Energy consumption in urban areas, much like food consumption, can be far removed from where environmental impacts occur, with populations remaining oblivious of the greenhouse gas and water pollution impacts of their consumption.

Due to the links between them, it is difficult to reliably project rates of spatial expansion in urban areas without accurate projections of population growth and GDP. The challenge is magnified by recent research suggesting that the relationship between these three factors can vary significantly across regions. Assessing changing urban spatial spread using satellites shows urban areas to be growing at an average rate of 3–7 per cent per year, with China exhibiting the highest rates. The contribution of population and GDP growth to this expansion has been found to be 28 and 72 per cent respectively for North America and 23 and 30 per cent respectively for India. In the same study, African city growth showed no relationship to GDP, although there is a recognition that in many developing countries there is significant informal economic activity that is not captured by GDP statistics.

In terms of the spatial distribution of people in growing cities, the defning feature, perhaps most common in East Asia, is peripheral development. Quantifying this phenomenon using satellite imagery for 2000 shows a range of estimates of the total spatial spread of urban areas of 0.2–2.4 per cent of the terrestrial land surface, due partially to difering defnitions of urban land cover. In developed countries such as the United States and Canada, about half the urban population lives in suburbs, while in the developing world squatter settlements or slums host more than one-third of urban populations.

The spatial distribution of cities demonstrates the complex interactions between urbanization and transport. For instance, when comparing per-person greenhouse gas emissions, Bangkok is dominated by transport emissions, while New York and London have significantly larger contributions from residential and commercial buildings. The ability to travel within a city is extremely important both in terms of the environmental impact and of economic productivity. In developing countries the majority of trips are taken by commuters but, as incomes increase, individuals are likely to make more personal trips. This preference often precipitates the acquisition of personal vehicles, as the locations of shopping or entertainment centres, schools or hospitals are widely spread and less easily connected by a public transport system. Finally, the type of fuel used is an important factor affeing the environmental impact of urban areas. Many trains already run on electricity, but should electric vehicle use increase, more electricity will be needed and – unless energy sources are priced according to their carbon intensity – an increase in electricity production using coal is likely, leading to significant increases in greenhouse gas emissions.

Cities have been seen as an opportunity for developing more sustainable resource management and reducing greenhouse gas emissions. While per-person emissions are generally lower in the cities of developed countries than in surrounding rural areas, the sources are much more diffuse and therefore difficult to manage with one overarching policy tool. Beyond mitigation activities, cities, particularly in developing countries, need to evolve climate adaptation measures. Several cities across South America, Africa and Asia have shown significant leadership in developing innovative adaptation strategies.

Developing cities are being encouraged to achieve zero waste, the principles of which include a reduction in waste incineration, the recycling of greater volumes of paper and plastics and the mining of precious metals and rare earth elements from existing landflls.

The question remains whether the Earth can support several billion additional people with a direct impact on land through subsistence farming, or additional urban billions with indirect impacts through consumer demand for fats and proteins from meats that are mostly produced on large corporate farms. The answer to this question will ultimately reveal how much land will be converted to livestock rearing, feedstock production and agriculture. Not evident in the short term is whether an accelerated or delayed demographic transition is more or less taxing on land systems. But if the living standards of the poorest are raised to more equitably match those of the developed world, then population growth should slow and the related environmental impact should begin to diminish.

Demographic and health transitions will continue to be major predictors of environmental change in general and of land-use and land-cover change in particular. Fundamental to facilitating demographic and health transitions will be investments in maternal and child health and education.

Globalization

Trade in food, fuels and minerals has increased dramatically over recent decades and shows few signs of slowing. International trade has grown rapidly since 1990, by 12 per cent per year, doubling in six years. In addition, annual emissions from exports have grown at 4.3 per cent, often due to production moving from developed countries to sites with less sophisticated technology in developing countries.

Greater liberalization of trade can exert pressure on the environment in any of three ways:

- increasing economic activity and by extension natural resource extraction, a scale effe;
- changing the type of economic activity to either more or less polluting industries, affeing intensity; and
- changing the technology or intensity of production that can sometimes encourage more environmentally friendly production techniques.

Regardless of the nature of the local change, wider trade allows the environmental impacts of production to be completely removed, or decoupled, from the site of consumption.

Such decoupling means that household consumption in developed countries can have significant environmental impacts elsewhere, particularly in developing nations. Tracing the impacts of consumption in Norway, Peters and Hertwich found that a household's environmental impacts in foreign countries embodied 61 per cent of its indirect emissions of CO_2, 87 per cent of sulphur dioxide, and 34 per cent of nitrogen oxides, while imports only represented 22 per cent of household expenses.

China is an instructive case for understanding trade. In the second half of the 20th century, it rapidly shifted its economy towards a processing base, resulting in a change from being a net exporter of primary resources to a net importer. Much of this processed merchandise is exported directly, with China's environment absorbing the pollution. Between 2002 and 2007, for example, 8–12 per cent of China's CO_2 emissions were attributable to exports to the United States.

Globalization is confounding the expected effect of the environmental Kuznets curve in countries with emerging economies. With afuence should come improvement in environmental conditions, but the link is proving difficult to confrm. In the case of China, nitrogen oxides and sulphur dioxide emissions have shown a complicated relationship with increasing income, suggesting that the reliance on coal-fred power may be negating improvements in other manufacturing technology.

Some invoke a traditional economic dynamic at work – a regulatory race to the bottom, where deregulation is expected to attract economic activity and create a comparative advantage over competitors. This notion suggests that concern for the environment and increasing environmental regulation in the developed countries result in migration of the most polluting industries to less afuent nations, although explicit evidence of this is inconclusive. A different explanation has also been ofered – that the pattern is more akin to the rapidly industrializing countries being stuck at the bottom, since there were no regulations to begin with. A related argument has also been made over the environmental effes of trade.

Either way, the consequence is the same – the creation of centres of pollution in developing countries. This suggests that the environmental Kuznets curve, relevant to a national context, has been disguising the displacement of pollution across national borders, with consumption in the most afuent nations driving environmentally polluting production and consumption to less afuent ones. For example, Cole (2006, 2004, 2003) has shown that trade increases environmental damage in the least developed countries while decreasing many forms of pollution in developed ones. Perhaps the environmental Kuznets curve does not work when all borders have been crossed by pollution.

Energy consumption and greenhouse gas emissions seem to follow this displacement pattern. A low-income country with less stringent regulations will find that an increase in trade openness increases energy consumption as its comparative advantage in dirty production deepens, while a high-income country will see energy consumption fall in response to trade liberalization.

So, will future goods produced for consumption inevitably also produce more pollution, despite regulations in developed countries? Carbon-intensive industries are leaving areas of stricter carbon regulation and moving to those that do not have such regulations. At the beginning of the 21st century, developed countries remained the largest greenhouse gas emitters in per-person terms. However, in the next few decades, the growth of emissions will come primarily from developing countries. So, despite 20 years of negotiations to avoid this outcome, developing countries will be following the same energy- and carbon-intensive development path as their developed counterparts have done.

Interactiion of Drivers

Drivers are interacting in unpredictable ways, resulting in some surprising consequences. This section, which links the drivers with a number of pressures on the environment, is intended to illustrate the complexity and provide some methods with which policy makers might be able to work to ameliorate the effes.

Critical Thresholds

Critical thresholds are being approached or even crossed. Ecosystems and the biosphere are systems that may change in a direct and linear way as a result of human stresses, or that may have more complicated dynamics. Although some can absorb a substantial amount of stress before they exhibit any response, change can take place abruptly and irrevocably when a threshold is exceeded, leaving little opportunity for human adaptation.

To understand the dynamics of a complex system, analysts seek out leverage points. The study of leverage points in complex systems suggests that indirect interventions can have great power and direct interventions can be used to enhance co-benefits, that both probable and possible outcomes should be addressed, and that difficult challenges can be broken down to manageable portions. The system must be monitored for both intended and unintended change.

The idea that the perturbation of a complex ecological system can trigger sudden feedbacks is not new: significant scientifc research has explored thresholds and tipping points that the planetary system may face if humanity does not control carbon emissions. From the perspective of drivers, understanding feedbacks reveals that many of them interact in unpredictable ways. Generally, the rates of change in these drivers are not monitored or controlled, and so it is not possible to predict or even perceive the thresholds as they approach. Critically, the bulk of research has been on understanding the effes of drivers on ecosystems, not on the effes of changed ecosystems on the drivers – the feedback loop.

In fact, Costanza *et al.* argues that this "great acceleration" began after the Second World War, with the scale of population growth and economic consumption and production increasing at rates that are orders of magnitude greater than in previous eras. It is this scale and speed that makes redirecting humanity's trajectory toward more sustainable development within the limits of planetary boundaries an extremely daunting challenge but one that we cannot aford to delay.

Overexploitation of Natural Resources

Considering that 14–16 per cent of animal protein consumed globally comes from the sea, overfishing of marine resources ofers a useful example of overexploitation of natural resources. At the global level, overfishing has been widespread but far from universal, and in those parts of the world with the capacity to manage fisheries, there is evidence that overfishing can be stopped and that previously overfished stocks can recover. There remain, however, a number of cases where overfishing continues despite the efforts of the international community, emphasizing the need for capacity building for both policy formulation and effective management.

The greatest expansion in fishing feets and harvesting occurred after the Second World War, as governments provided significant subsidies to encourage more investment in harvesting technologies, which massively increased yields. In many cases the increased yield proved to be unsustainable, and fishery declines were widespread by the 1970s. Extension of jurisdiction with the United Nations Convention on the Law of the Sea (UNCLOS) resulted in improved management practices in many coastal areas, but a second round of expansion of fishing capacity resulted in a second round of declines. Overcapacity remains a serious problem in global fisheries despite an international agreement to address it, the 1999 International Plan of Action for the Management of Fishing Capacity.

Part of the problem in sustainably managing fisheries is the difficulty of monitoring the state of fish populations, especially in areas outside the jurisdiction of national or international authorities where biological information and even basic catch data may be unavailable or unreliable. Moreover in many fisheries, data are not recorded on species taken as by-catch – unwanted fish caught inadvertently, often returned to the sea dead or dying – so their status and the impacts of fishing are unknown and unmanaged. More generally, poor monitoring means that there is little knowledge about the dynamics of many fish populations, making it difficult to discern whether the observed populations are showing signs of natural variability or imminent collapse.

Driver Combinations and Feedbacks on Human Health

Looking specifically at food production, human and ecosystem exposure to chemicals increased dramatically with the industrialization of agriculture. There has been limited research on the human and environmental health impacts of long-term exposure to these chemicals, but it is known that the risks are much higher in developing countries where 99 per cent of current global deaths from pesticide exposure occur, both from occupational exposure and from casual exposure resulting from lax or absent health and safety controls.

Nitrate pollution from both crop cultivation and livestock production is among the most destructive impacts of food production, with the scale of meat production having serious ramifications for local pollution levels. In the United States, for example, of the top 20 sources of industrial pollution, eight are slaughterhouses. In addition, the country's Concentrated Animal Feeding Operations (CAFOs) produced 500 million tonnes of manure in 2007: three times the United States' 2007 total amount of human waste. A further problem from centralized meat production facilities involves how bacteria convert excess nitrate in such waste into nitrous oxide, a potent greenhouse gas, or it can leach into waterways and groundwater.

Generating Intense Pressures

Drivers of environmental change are growing, evolving and combining at such an accelerating pace, at such a large scale and with such widespread reach that they are exerting unprecedented pressure on the environment. Most forms of consumption and production use the environment as a source of raw materials and as a sink for wastes. The impacts can be highly concentrated in some parts of the world – such as nuclear waste storage facilities and residual accumulation of toxic compounds at e-waste recycling sites – or systemically spread over the entire globe – such as PCBs delivered along the food chain from equator to poles – and they can quickly create new and potentially dangerous situations. In many instances their impacts can be so deep, rapid and unpredictable that they risk exceeding environmental thresholds and societal capacity to monitor them or respond adequately.

The combination and scale of some drivers can create dynamic patterns that, in turn, generate complex systemic interactions. One example is the rise in greenhouse gas emissions, the scale of which has defied global efforts to stimulate the necessary action to stem emissions. In addition to rising global temperatures and sea levels, scientists predict that the pace and scale of climate change could eventually exceed certain ecological limits or thresholds, leading to surprising and dangerous consequences such as the alteration of the world ocean's chemical composition with increasing proportions of acidifying carbon, the global loss of coral reef ecosystems, or the collapse of the West Antarctic ice sheet.

One driver can trigger a series of drivers and pressures that act in a domino fashion. For example, concerns about climate change impacts, including crop vulnerability and food insecurity, gave rise to policies that included mandates to increase biofuel production, such as legislation introduced in 2003 in the EU and in 2008 in the United States. The resulting demand generated a cascading set of pressures including crop diversion to biofuels. This diversion of cropland then contributed to higher food prices in 2008 and 2010, increasing worries about food insecurity.

Inertia and Path Dependencies

As global ecological and institutional systems are extremely complex and slow to change, decisions made today have long-term and far-reaching impacts. Without addressing the drivers behind the current trajectory, it will be difficult to move to an environmentally sustainable suite of choices and outcomes. At the same time the need for urgency must be recognized. Finally, due to the inertia in the system and an unwillingness to address these drivers in the past, future generations are committed to a range of impacts that could have been avoided. The most daunting of these problems is climate change, where a coalescing of several drivers has made reducing carbon emissions a very complicated task. For instance, current fossil-fuel-dependent energy and transport infrastructures are estimated to have committed the planet to emitting 496 billion tonnes of CO_2 from now until 2060. These calculations do not include currently uncommitted transport network extensions, additional fossil-fuel-based power plants or the complex economy of refuelling stations or factories dependent on combustion energy, all of which are entirely reliant on the current model of energy generation and transport. The issue is not solely about the existing physical infrastructure that would be costly to replace, but the millions of jobs, processing facilities and entire sub-industries that have developed as a result of the *status quo*.

The case for investments in transport infrastructure has been made before. However, the institutionalization of global food production ofers similar barriers to change. United States farm policy provides an illustrative example of this phenomenon, although it is by no means the only country where it occurs. Currently, 74 per cent of agricultural land in the United States is dedicated to eight commodity crops: maize, wheat, cotton, soybeans, rice, barley, oats and sorghum, supported by 70–80 per cent of government agricultural subsidies, while the farming industry has consolidated to become an industrialized food production system. Unfortunately, the emphasis on producing these eight crop commodities has resulted in a food system where healthier food options, such as vegetables and fruits, increased in price by more than 100 per cent between 1985 and 2000, while the price of unhealthy fats and oils derived from these basic foodstufs rose by only 35 per cent. With many of the country's consumers making daily consumption decisions based on cost, decades of investment in this vertically integrated and politically powerful industry make fundamental changes in the health outcomes of the food system extremely challenging.

However, not all health effes are diet related, but can be linked to such atmospheric pollution as nitrate formation and chemical pollution resulting from enhanced pesticide use, amongst other sources. For instance, in the United States, a high proportion of maize and soybean crops are genetically modified to resist the effes of the herbicide glyphosate, applied in vast quantities to eradicate weeds. Within the supply chain, maize and soy make up 83–91 per cent of livestock feed grains. Ongoing research raises the question of the endocrine-disrupting potential of glyphosate. The residence time of glyphosate in the environment is difficult to model, as it is dependent on a number of biophysical factors and monitoring capability is only recently catching up with its widespread use. However, in communities located near agricultural felds, evidence of glyphosate and its most common degradate aminomethylphosphonic acid (AMPA) can be found in the atmosphere, rain and local water bodies.

References

Bongaarts, J. (2001). *Household Size and Composition in the Developing World.* Population Council, New York

Carr, D. (2009). Population and deforestation: why rural migration matters. *Progress in Human Geography* 33(3), 355378

Dietz, T., Fitzgerald, A. and Shwom, R. (2005). Environmental values. *Annual Review of Environment and Resources* 30, 335372

Enerdata (2011). *Global Energy Statistical Yearbook.* Enerdata, Grenoble

Peters, G.P. and Hertwich, E.G. (2006). The importance of import for household environmental impacts. *Journal of Industrial Ecology* 10(3), 89110

Seto, K.C., Sanchez-Rodriguez, R. and Fragkias, M. (2010). The new geography of contemporary urbanization and the environment. *Annual Review of Environment and Resources* 35, 167194

2

Global Atmospheric Governance

Substances emitted to the atmosphere as a result of human activities are a challenge to both the environment and development: millions of people die prematurely each year from indoor and outdoor air pollution; ozone-depleting substances (ODS) have thinned the ozone layer and created seasonal holes in the stratospheric ozone layer over polar regions; and climate change is happening now, and atmospheric concentrations of greenhouse gases and other substances that affect climate continue to increase. Climate change threatens, amongst other things, food security and biodiversity, and it is likely to increase storm damage on all parts of the globe. People in many of the developing regions are especially vulnerable.

These atmospheric issues are addressed by several global and regional agreements including Agenda 21 and the Johannesburg Plan of Implementation. Internationally agreed goals and, in some cases, targets have been established. In addition, there are some internationally agreed guidelines related to human health and ecosystems that are used to monitor progress in addressing atmospheric issues.

It uses key indicators to assess progress in relation to goals set at global and regional levels for atmospheric issues. It considers whether progress is on track to achieve these goals using existing policies and measures, and whether they are sufcient to address the key issues important to human well-being and development. It then considers the outlook for the different issues and what more needs to be done.

The scientifc basis for the development of air pollution policy has greatly improved and there is increasing understanding of the socio-economic aspects of atmospheric issues. Recently, science has pointed to new challenges such as near-term climate change and short-lived climate forcers (SLCFs), and knowledge about thresholds and tipping points has improved.

Climate change, air quality and stratospheric ozone depletion are closely related, as individual pollutants can have multiple impacts on health, crop yields, ecosystems, cooling or heating of the atmosphere and stratospheric ozone depletion, all with the potential to affect human well-being. Many sources also emit multiple pollutants that can both affect air quality and cause climate change. Yet, despite these links, most governments address these issues separately, in part because goals were set in this way 20 years ago. Depending on which measures are implemented, there could be co-benefcial or antagonistic outcomes and, unless a more integrated approach is developed, there is a risk that different atmospheric policies could work against each other.

International Goals and Targets

Major goals to protect the environment and human well-being from the impact of substances emitted to the atmosphere were established in Agenda 21 and the Johannesburg Plan of Implementation. These emphasized the need to identify threshold levels of pollutants and greenhouse gases that cause "dangerous anthropogenic interference with the climatic system and environment". Meeting the objectives to phase out chlorofuorocarbons (CFCs) and other ozone-depleting substances – as defned in the 1985 Vienna Convention for the Protection of the Ozone Layer and its 1987 Montreal Protocol on Substances that Deplete the Ozone Layer – was considered essential. These also recognized the importance of the 1979 Convention on Long-Range Transboundary Air Pollution (CLRTAP) and its protocols to reduce regional air pollution, and recommended that these programmes be continued and enhanced, and their experience shared with other regions.

The Johannesburg Plan of Implementation went on to consider air quality as a part of overall development, promoting an integrated approach to policy making. It stressed the need to reduce respiratory diseases and other health impacts resulting from air pollution, paying particular attention to women and children. It supported the phasing out of lead in petrol, measures to prevent children's exposure to lead, and efforts to strengthen the monitoring, surveillance and treatment of lead poisoning. Another focus was to assist developing countries in providing afordable energy to rural communities, particularly to reduce dependence on traditional fuels for cooking and heating.

Other non-atmosphere-related conventions such as the 1992 Convention on Biological Diversity (CBD) also have links with the impacts of atmospheric pollution. The Aichi Biodiversity Targets include two atmosphere-related targets:

- *Target 8:* by 2020, pollution, including from excess nutrients, will have been brought to levels that are not detrimental to ecosystem function and biodiversity; and
- *Target 10:* by 2015, the multiple anthropogenic pressures on coral reefs, and other vulnerable ecosystems impacted by climate change or ocean acidification will have been minimized, so as to maintain their integrity and functioning.

Atmospheric goals and targets are supported by both legally and non-legally binding environmental agreements, most of which contain globally agreed quantitative targets and timelines for implementation that have catalysed the development and implementation of national regulation. The goals and targets refer to different aspects of control, including:

- control of drivers, for example the total ban - with a few exceptions - of the production and consumption of ozone-depleting substances, and the phase-out of leaded petrol;
- reducing pressures, for example emission reductions of carbon dioxide (CO_2) and other greenhouse gases; and
- targeting concentrations of, for example, particulate matter (PM) and CO2.

For outdoor and indoor air pollution there are no global targets as such, but the World Health Organization (WHO) has established air quality guidelines, based on scientifc research, to help assess progress towards reducing risks from air pollution. The limit for global temperature increase at the end of the century – the agreed 2°C limit – was set on the basis of scientifc discussion of the potential impacts, but also political realities and the likelihood that it could be achieved. Countries set national air-quality standards and even greenhouse gas commitments or targets specifc to their international obligations, development situation and institutional capacities. The Copenhagen Accord invited developed countries to submit economy-wide emission reduction targets for 2020 and developing countries to submit nationally appropriate mitigation actions (NAMAs). The Cancun Agreements legally recognized these pledged targets and actions, formally anchoring them in the United Nations Framework Convention on Climate Change (UNFCCC). The CLRTAP remains the only regional agreement on transboundary air pollution that sets targets for many different pollutants. Some regions, and sub-regions – Africa, Asia and South America – have cooperation agreements that show intent to reduce emissions, but these are not binding, and in some cases have not been implemented due to lack of human and financial resources.

Progress in Achieving Atmospheric Goals

The goals and targets set globally and regionally for a number of atmospheric issues are compared to the current situation, examining whether they have been met and determining the size of the gap between the current situation and the goals and targets.

Progress is described against key indicators by considering atmospheric issues in three main categories:

- examples of where targets are not being met and the situation remains far from sustainable;
- examples of mixed progress, with some regions having met targets and others remaining far from them; and
- examples of good progress, where targets have been set and are largely met.

Goals Far from being Met

There is broad scientifc consensus that anthropogenic emissions of CO_2 and other greenhouse gases are the leading cause of contemporary climatic changes. Four independent analyses show that 2000-2009 was the warmest decade on record with atmospheric concentrations of CO_2 also increasing. A look at regional temperature changes shows that the greatest warming over the past century is at high latitudes.

Climate change threatens human well-being in many ways, from a greater frequency of heat waves and severe storms, to shifts in rainfall patterns and rising sea levels. Changes in the frequency of tropical cyclones are uncertain, but it is likely that their intensity will increase with rising temperatures. There is growing concern that inaction will lead to changes that are irreversible at human timescales – so-called tipping points. Increased release of carbon stored in permafrost, as CO_2 or methane, is an example of a change that could give rise to a cycle of further warming and further releases of greenhouse gases.

Concentrations and emissions of most anthropogenic greenhouse gases have increased during recent years. Growth rates have been especially high for concentrations of several hydrofuorocarbons (HFCs) while emissions of CO_2 from fossil fuel consumption have followed the more pessimistic of the widely used projections of the Intergovernmental Panel on Climate Change *Special Report on Emission Scenarios* (SRES) during the last decade, despite a brief downturn in global emissions in 2009 associated with the economic recession. The rapid growth of CO_2 concentrations is also associated with similarly rapid increases in ocean acidification.

To avoid exceeding the 450 ppm atmospheric concentrations of CO_2-equivalent that are likely to be required to stay within the temperature rise limit of 2°C, the IPCC has concluded that developed countries need to reduce emissions by 25–40 per cent below 1990 levels by 2020, while peer-reviewed literature has concluded that developing countries need to reduce emissions by 15–30 per cent relative to business-as-usual by 2020. Further reductions are then required beyond 2020 to achieve the target. While some countries have reduced CO_2 emissions since the Kyoto Protocol entered into force in 2005, many appear unlikely to reach their Kyoto targets. Further, many of the same countries reporting reductions have increased imports of carbon-intensive products – so-called carbon leakage. Accounting for the foreign CO_2 emissions embedded within imported products, emissions have in fact increased in many developed nations and the net domestic-plus-embedded emissions are far greater than the Kyoto targets.

The years since the Bali Action Plan have seen 42 developed countries pledge quantified economy-wide emission targets up to 2020, while 44 developing countries have pledged nationally appropriate mitigation actions. The gap between expected emissions and the agreed UNFCCC 2°C limit lies between 6 billion and 11 billion tonnes of CO_2-equivalent. The size of the gap depends on the extent to which the pledges are implemented and how they are applied.

Table 1. Concentrations of greenhouse gases, 2005, 2009 and 2010

	2005	*2009*	*2010*
CO_2 (ppm)	378.7	386.3	388.5
CH, (ppb)	1 774.5	1 794.2	1 799.1
NLO (ppb)	319.2	322.5	323.1
CFC-11 (ppt)	251.5	243.1	240.5
CFC-12 (ppt)	541.5	532.6	530.8
HCFC-22 (ppt)	168.3	198.4	206.2
HFC-134a (ppt)	34.4	52.4	57.8

An historical tendency to underestimate rates of climate change suggests that non-linear changes and material losses at the higher end of estimated ranges are also possible. Overall, prospects for long-term climate change look bleak if there is no demonstrable progress at both international and national levels.

Even if negotiations take longer than expected at the international level, national actions should continue to move forward. A growing body of low-carbon research has shown that in countries ranging from the United Kingdom and Japan to Thailand, it would be economically and technically feasible to cut emissions in half by 2050. The results of these studies are based on placing a price on carbon through, for instance, an emissions trading scheme. It is important to note, however, that market-based instruments such as emissions trading schemes or the Clean Development Mechanism (CDM) may not work in all contexts or benefit all regions equally. For example, in the CDM market, Latin America and Asia and the Pacifc account for more than 87 per cent of all projects while Africa accounts for less than 3 per cent.

Other studies suggest that mainstreaming climate change into existing development plans could provide a more promising alternative to market-based instruments, especially for those developing countries that constrain their development with an emissions cap. This is reinforced by research that shows that because of the relatively greater value of co-benefits, such as improved local air quality, low-income countries have the most to gain from mitigating greenhouse gases in a manner consistent with development priorities. Capturing these co-benefits not only requires that policy makers become adept at mainstreaming climate change in development plans, but also necessitates decision-making frameworks that explicitly acknowledge synergies between climate change and other atmospheric issues. Such an integrated approach can readily be put in place at local and city levels, where a considerable amount of climate mitigation and air quality control has already been implemented.

Mixed Progress

There are examples of improvement in some regions while large difficulties remain in many others, and global targets are still far from being met. Four major atmospheric issues are described below: sulphur, nitrogen, small particulate matter (typically described as PM_{10} and $PM_{2.5}$) and tropospheric ozone.

Sulphur Pollution

Sulphur dioxide (SO_2) emissions, predominantly from fossil fuel use in power generation, industry and transport, have detrimental effes on human health by contributing to $PM_{2.5}$, on terrestrial and freshwater ecosystems by acidification, on man-made materials and cultural heritage by corrosion, and on biodiversity and forestry. Sulphate aerosols also cool the atmosphere, which makes it important to track them in order to assess the overall benefits of greenhouse gas reduction strategies.

Since issues of transboundary air pollution were highlighted in Agenda 21 there have been considerable reductions in sulphur dioxide emissions in Europe and North America, achieving the targets of the CLRTAP protocols, the National Emission Ceiling (NEC) Directives of the European Union (EU) and clean air legislation in Canada and the United States. Key to the development of country targets in Europe was the use of critical loads (deposition thresholds above which harmful effes are observed). Successful implementation of legislation brought about a drop of around 20 per cent in global emissions between 1980 and 2000.

Emissions from Europe and North America were dominant until about 2000, when East Asian emissions started to dominate. According to the Representative Concentration Pathway (RCP) scenarios, global sulphur dioxide emissions were projected to decline steadily after 2005, and by 2050 to be 30, 50 or 70 per cent lower than 2000 levels. This set of four new pathways was developed for the climate modelling community as a basis for near- and long-term modelling experiments.

As sulphur deposition has abated in Europe and North America, acidification has also diminished and some freshwater ecosystems have recovered, although critical loads are still exceeded in some areas. In Asia, the increase in emissions has put sensitive ecosystems at risk from the effes of soil acidification. However, the large-scale acidification of natural lakes experienced in Europe and North America has not been seen in Asia, and may be unlikely due to the nature of the region's soil and geology. In 2005 it was estimated that the critical load for soils in China was exceeded by sulphur deposition across 28 per cent of the country's territory, mainly in eastern and south-central China. The area in exceedance is projected to decrease to 20 per cent in 2020, given the implementation of current plans for emission reductions.

Further action on sulphur emissions is being taken through the revision of the Gothenburg Protocol in Europe. In Asia, action is also being taken to improve the efciency of energy use and reduce sulphur dioxide emissions. For example, as part of its five-year plans, China implemented fue-gas desulphurization and the phasing out of small, inefcient units in the power sector in a move to achieve the national goal of a 10 per cent reduction in sulphur dioxide emissions between 2005 and 2010.

Global efforts are also being made to reduce sulphur emissions from key sectors, including transport and shipping. The human health effes of particulate matter measuring 2.5 micrometres or less in diameter ($PM_{2.5}$) are being tackled by lowering the sulphur content of diesel fuels – for example, UNEP's Partnership for Clean Fuels and Vehicles (PCFV) is promoting the reduction of sulphur in vehicle fuels to 50ppm or below worldwide. Sulphur emissions from shipping have become an important policy issue in Europe, while the International Convention for the Prevention of Marine Pollution from Ships (MARPOL) is committed to the progressive global reduction in emissions of sulphur oxides, nitrogen oxides and particulate matter.

Nitrogen Compounds

Human activity linked to energy use and food production has more than doubled the amount of reactive nitrogen circulating in the environment over the past century. This is emitted to the atmosphere as nitrogen oxides (NO_X), mainly from the transport and industry sectors, and ammonia (NH_3) and nitrous oxide (N_2O), mainly from agriculture. They have multiple effes on the atmosphere, terrestrial ecosystems, freshwater and marine systems, and on human health, a phenomenon known as the nitrogen cascade. Nitrogen compounds are precursors of atmospheric $PM_{2.5}$, which has impacts on human health, while nitrogen oxide is a precursor of tropospheric ozone, which has impacts on health, crop yields, ecosystems and climate. Nitrous oxide and tropospheric ozone are also important greenhouse gases. Nitrogen deposition drives biodiversity loss through eutrophication and acidification in terrestrial and aquatic ecosystems. However, it can also be of benefit to crop yields, and can increase carbon sequestration through the stimulation of forest growth.

Total global nitrogen oxide emissions increased until around 2000, but were expected to remain more or less constant thereafter, with reductions in Europe and North America compensating for the growth in emissions in Asia and all other regions. Control measures in Europe – where road transport accounted for 40 per cent of emissions in 2005 – succeeded in reducing total nitrogen oxide emissions by 32 per cent between 1990 and 2005, while measures in the United States reduced emissions by 36 per cent between 1990 and 2008. In Asia, emissions have continued to increase over the past two decades, with the growth rate itself accelerating during this period. Emissions from international shipping are estimated to have risen from 16 million tonnes in 2000 to 20 million tonnes of nitrogen dioxide (NO_2) in 2007.

Global ammonia emissions, largely from the agricultural sector, have increased fivefold since the middle of the last century and are projected to continue to climb in all regions with the possible exception of Europe, where they have decreased slightly and may stabilize. Nevertheless, there is a lack of concern and focus on this issue in Europe, and there is often resistance to major changes from the farming community. In most other regions ammonia is not regulated under major emission control laws. However, the Gothenburg Protocol of CLRTAP is being revised with more stringent targets and is likely to lead to a further reduction in emissions in Europe.

Despite these improvements, nitrogen-based air pollution from agriculture, industry and trafc in urban areas contributes significantly to $PM_{2.5}$ concentrations as secondary nitrate and ammonium particles, which are reducing people's life expectancy by several months across much of Central Europe.

In Africa, Asia and Latin America, where control of nitrogen emissions is not a high priority, projections show increases in emissions of both nitrogen oxides and ammonia. In some regions, especially in Africa, lack of monitoring capacity is a major issue. To address this, more policy emphasis on these substances in these regions will be required, especially with regard to emissions from the agricultural, energy, industrial and transport sectors, while ensuring that there is adequate nitrogen fertilizer available for food production.

Current technology can deliver significant reductions in emissions of nitrogen oxide, but growth in certain sectors, particularly transport, can counteract control measures. Changed management practices will be needed to reduce ammonia emissions, and more fundamental consideration of agricultural policy and practice, as well as changes in consumption patterns of meat and dairy products, are required if large reductions are to be achieved.

Rising atmospheric nitrogen deposition will lead to environmental effes associated with the nitrogen cascade, including impacts on plant diversity. The Convention on Biological Diversity has recognized nitrogen deposition as an indicator of the threat to biodiversity and especially to sensitive ecosystems that receive a total nitrogen deposition above 10 kg per hectare per year. However, the full impact is difficult to estimate as there is little quantification of the effes on biodiversity outside Europe and North America.

Designing effective policies to balance the positive impacts of nitrogen deposition, such as increases in crop yields and carbon sequestration, and the negative impacts, such as loss of biodiversity and increased greenhouse gas emissions, underlines the necessity for a truly integrated approach to nitrogen management in the environment.

Particulate Matter

Control of particulate matter has achieved mixed progress worldwide. In Europe and North America, as well as some cities in Latin America and Asia, emissions of PM_{10} – particles of 10 micrometres in diameter or less – have been reduced, but they remain a major pollutant in many other cities in Asia and Latin America. Very few cities in Africa monitor air pollutants; however, of the few that do, many show PM_{10} concentrations in excess of WHO guidelines. Outdoor concentrations in high-income countries come close to the WHO PM_{10} guideline of 20 micrograms per m^3. In Africa, the most widespread issue is indoor levels of particulates. Regulating these pollutants is complex because they are composed of a variable mix.

Global Trends

Mixed progress in relation to WHO guidelines, with significant reductions in the EU and North America and some Latin American and Asian cities, but mostly high concentrations in urban areas in Asia and Latin America; data for Africa is insufficient, but some cities have high PM levels of primary emissions and secondary pollutants, where the original emissions are transformed in the atmosphere. An additional challenge for cities is the elimination of particulate hotspots.

Particulate matter, especially the fner $PM_{2.5}$, is the most important air pollutant causing damage to human health. The prime sources of particulate matter relate to the energy, transport and industry sectors, but open burning of solid waste and crop residues are also important sources. Health research worldwide has shown that there is no safe threshold for exposure, as even very low levels cause health damage. Impacts on health are predominantly associated with respiratory and cardiovascular illnesses, but the range of effes is broad for both acute and chronic exposure. Based on exposures to particulate matter in 2004, WHO estimated that annually 5.3 per cent of premature deaths worldwide, about 3.1 million people, are attributable to air pollution – 2 per cent to outdoor urban pollution and 3.3 per cent to indoor pollution – which is more than from all other environmental risks combined. However, a more recent study estimated 3.7 million premature deaths due to outdoor anthropogenic $PM_{2.5}$ alone, as it used a different method that includes exposure in rural areas, does not have a low-concentration threshold, and used updated concentration-response relationships. Worldwide, approximately 41 million disability-adjusted life years (DALYs) – the sum of potential healthy life years lost due to illness – are attributed to solid fuel and methods use with about 18 million, or 44 per cent of the total, occurring in sub-Saharan Africa. Household energy interventions, which reduce dependence on traditional fuels and methods for cooking and heating, clearly have the potential to improve health and promote achievement of the MDGs. Even in high-income countries such as the United Kingdom, $PM_{2.5}$ is estimated to have contributed to 29 000 premature deaths and the loss of 340 000 life-years in 2008, despite considerable progress in reducing concentrations.

Recent assessments of the long-range transport of air pollution indicate that intercontinental transport of particulate matter is contributing to exceedances of public health standards and visibility targets. Long-range transport of particulates may be responsible for 380 000 premature deaths worldwide, of which 75 per cent are attributable to (mineral) dust $PM_{2.5}$.

Various measures, including technological improvements to vehicles, increased transport and energy efciencies and cleaner fuels and flters, have been successful in developed countries and to some extent in developing ones. However, while the latter are catching up in the use of cleaner technologies, such efciencies are being compromised by a rapid increase in emission sources, for example fuel use for energy and transport. Where indoor particulates are concerned, global partnerships are promoting cleaner energy and improved cooking stoves.

Most developed and developing countries have adopted ambient air quality standards, but concentrations of particulates in most cities exceed the levels recommended by WHO's ambient air quality guidelines for protecting human health and ecosystems. Most of the PM_{10} standards in developing countries are less stringent than the interim targets set by WHO to promote a progressive reduction in air pollution. WHO has also recommended $PM_{2.5}$ guidelines, but many countries have yet to adopt standards and monitoring practices. In Asia in 2010, for example, only four out of the 22 countries have standards for $PM_{2.5}$ supported by monitoring.

Projections of a 20 per cent reduction in $PM_{2.5}$ emissions by 2020 in Europe are expected to lead to a 40 per cent fall in the associated years of life lost compared to the year 2000; nonetheless, $PM_{2.5}$ air pollution is still expected to shorten statistical life expectancy by 4.6 months. However, if the new National Emission Ceilings in Europe are implemented the benefits will outweigh the costs 12–37 times depending on the valuation method, and PM emissions could be reduced by 35–50 per cent depending on the portfolio of measures.

There are some uncertainties that need to be resolved to inform better policy making for particulate matter and health. These include the concentration and impact of particle sizes and a better understanding of the nature of primary and secondary PM pollution in different locations through monitoring, emission inventories and modelling, as well as through the use of source apportionment and estimation of the economic value of health impacts. Eforts to harmonize ambient air quality standards and the building of capacity have the potential to fast-track PM reduction in developing countries, amplifying the successful policies and technologies applied in Europe and North America and in some Asian and Latin American cities.

Tropospheric and Surface Ozone

Tropospheric ozone (O_3) in the lower atmosphere, from 0–10 up to 20 km above the Earth's surface, is responsible for ozone's impact on warming. Ground-level or surface ozone refers to concentrations at ground level that affect both human health and ecosystems. There is mixed progress in controlling tropospheric ozone: peak concentrations have decreased in Europe and North America, and North America, while background concentrations have increased. In rapidly industrializing regions both background and peak concentrations have been steadily rising.

Ozone causes harm in three main ways. Firstly, surface ozone damages human health and its impact is considered second only to particulate matter. It is responsible for an estimated 0.7 million respiratory deaths globally each year, more than 75 per cent of which are in Asia. Ozone can also have chronic health effes resulting in permanent lung damage.

Secondly, surface ozone is the most important air pollutant causing damage to vegetation, diminishing crop yields and forest productivity and altering net primary productivity. Estimates suggest, for example, that ozone-induced yield losses range between 3 and 16 per cent for four staple crops – maize, wheat, soybean and rice – which translate into annual global economic losses of US$14–26 billion.

Lastly, ozone is the third most important greenhouse gas after CO_2 and methane, but is classified as a short-lived climate forcer due to its residence time in the atmosphere of just days to weeks. Tropospheric ozone is estimated to have been responsible for a change in radiative forcing of +0.35 (-0.1, +0.3) watts per m^2 since pre-industrial times, compared to a combined anthropogenic radiative forcing of +1.6 (-1.0, +0.8) watts per m^2. These ozone-induced changes are thought to be responsible for 5–16 per cent of the global temperature change since pre-industrial times. Reductions in biomass caused by ozonealso influence the amount of carbon sequestered within terrestrial ecosystems. This effect is estimated to increase atmospheric CO_2 concentrations such that the additional radiative forcing could exceed warming due to the direct radiative effect of tropospheric ozone in the atmosphere.

Ozone is not directly emitted into the atmosphere but rather is formed when precursor pollutants – nitrogen oxides and volatile organic compounds, including methane, and carbon monoxide – react in the presence of sunlight.

As such, ozone concentrations tend to be higher at some distance – tens to thousands of kilometres – downwind of precursor pollutant sources, causing ozoneto pollute at the local, regional and hemispheric scale.

Photochemical reactions account for approximately 90 per cent of the ozone in the troposphere, with the remaining 10 per cent directly transported from the stratosphere. Around 30 per cent of tropospheric ozone is due to anthropogenic emissions, with 40 per cent of the change in the global ozone burden since pre-industrial times due to increases in methane, and the remainder to increases in emissions of nitrogen oxides, carbon monoxide and volatile organic compounds other than methane. The origin of ground-level or surface ozone relevant to effes on human health and ecosystems across the polluted regions of the northern hemisphere is 20–25 per cent from the stratosphere and a similar proportion from natural precursor sources including lightning and emissions from soils, vegetation and fre, along with oxidation of naturally occurring methane. The anthropogenic contribution thus typically exceeds 50 per cent over these regions.

Elevated ozone concentrations tend to be associated with regions experiencing high levels of uncontrolled emissions from industrial and urban centres as well as seasonal periods with high solar radiation. This causes high variability in global and seasonal concentrations. Regions in North America, Europe and Asia have been identified as having a high anthropogenic ozone load.

The targets defned for tropospheric ozone in the United Nations Economic Commission for Europe (UNECE) region are now being exceeded at many locations. Coordinated action in Europe, however, has resulted in emissions of nitrogen oxide and volatile organic compounds currently being 30 and 35 per cent lower than in 1990, leading to reductions in short-term peak ozone concentrations to daily peak values of around 60 micrograms per m^3 of air. In contrast, mean ozone concentrations at many locations have been increasing due to a variety of different factors. For example, local emission reductions of nitrogen oxides, and hence nitric oxide, remove a key mechanism of ozone destruction and can result in increases in concentrations in urban areas. There is also evidence of background ozone concentrations increasing by up to 10 micrograms per m^3 of air per decade since the 1970s due to changes in stratospheric ozone incursions, hemispheric transport and ozone formation responding to climate change. This will increase both mean and peak ozone levels.

Global photochemical modelling studies performed for the HTAP assessment provide estimates of changes in surface ozone concentration for those regions that currently show the highest concentrations. These data indicate recent reductions in surface ozone in North America and Europe, which are likely to be due to effective controls of nitrogen oxides and volatile organic compounds over the past two decades in response to the US Clean Air Act and CLRTAP and EU targets in Europe. In contrast, trends in Asia are continuing upwards due to continued rapid industrialization across the region. However, these regional trends may hide large local variations.

Future changes in tropospheric ozone concentrations have been explored using a number of different global photochemical models for a variety of emission scenarios and provide variable results. The HTAP assessment used a mean from six global models to assess the implications of emission changes between 2000 and 2050 following the RCP emission scenarios. The outlook for ozone concentrations is heavily dependent on global and regional emission pathways.

Assessment of the effectiveness of policies introduced to control ozone requires an extended global monitoring network covering rural as well as urban locations. Improved understanding of ozone'simpact on human health and ecosystems and how climate change will affect its formation, as well as how ozone acts in combination with other stressors such as global warming and excessive nitrogen deposition, will also be important. The growing interest in ozone as a short-lived climate forcer and the associated human health, arable-agriculture and ecosystem benefits that its reduction might bring make this a pollutant of particular interest for policy intervention.

Progress on Internationally agreed Goals

There are two examples of substantial progress in solving issues and achieving targets: protection of the stratospheric ozone layer and the removal of lead from petrol.

Stratospheric Ozone Layer

Global regimes to address stratospheric ozone depletion include the 1985 Vienna Convention for the Protection of the Ozone Layer and its 1987 Montreal Protocol on Substances that Deplete the Ozone Layer. The latest scientifc assessments confrm the success of the action taken under the Montreal Protocol to eliminate consumption of ozone-depleting substances.

Stratospheric ozone protects humans and other organisms because it absorbs ultraviolet-B (UV-B) radiation from the sun. In humans, heightened exposure to UV-B radiation increases the risk of skin cancer, cataracts and suppression of the immune system. Excessive UV-B exposure can also damage terrestrial plant life, single-cell organisms and aquatic ecosystems. In the mid-1970s, it was discovered that the thinning of the stratospheric ozone layer was linked to the steady increase of chlorofuorocarbons (CFCs) – used for refrigeration and air conditioning, foam blowing and industrial cleaning – in the atmosphere.

The most severe and surprising ozone loss – which came to be known as the ozone hole – was discovered to be recurring in springtime over the Antarctic. Thinning of the ozone layer has also been observed over other regions, such as the Arctic and northern and southern mid-latitudes.

The hole in the Antarctic ozone layer is the clearest manifestation of the effect of ozone-depleting substances: springtime Antarctic total column ozone losses continue to occur each year, with the extent affeed by meteorological conditions.

Model simulations for an ozone depletion scenario without the Montreal Protocol, the "World Avoided" scenario, show that there would have been a 300 per cent increase in the UV radiation at mid-northern latitudes or a 550 per cent increase in UV radiation at the mid-latitudes by 2065 compared to 1980 levels. Such a drastic increase in UV-radiation would have had serious consequences for both human health and the environment. In the United States alone, it is estimated that 22 million cases of cataract will be avoided for people born between 1985 and 2100 and 6.3 million skin cancer deaths will be avoided up to the year 2165 as a result of the Montreal Protocol.

The most recent amendment of the Montreal Protocol, in 2007, accelerated the phase-out of hydrochlorofuorocarbons (HCFCs), contributing to a reduction in global warming potential (GWP) of about 18 billion tonnes of CO_2-equivalent emissions.

The current phase-out of ozone-depleting substances is expected to lead to recovery of the ozone layer at different times in different regions. For the world as a whole, annually averaged total column ozone is projected to return to 1980 levels between 2025 and 2040, but this will take until mid-century in the Antarctic, with small episodic Antarctic ozone holes likely to persist even at the end of the 21st century. Annually averaged total column ozone is projected to return to 1980 values between 2015 and 2030 over northern mid-latitudes, while for southern mid-latitudes it is projected to recover between 2030 and 2040.

Despite the successful implementation of certain Montreal Protocol provisions, some issues remain regarding the capture of ozone-depleting substances in old equipment and the destruction of collected or stockpiled appliances.

Removal of Lead from Petrol

The goals in the Johannesburg Plan of Implementation to reduce exposure to lead have largely been met, with most countries having phased out lead in petrol since 2002, although there is evidence that leaded petrol is still sold in at least six countries.

Lead poisoning, at all levels of exposure, causes adverse and often irreversible health effes in humans, particularly children, and accounts for about 9 million DALYs or around 0.6 per cent of the global burden of disease. Acute exposure to high levels of lead affes the brain and central nervous system, causing coma, convulsions and even death. Lead can also adversely affect the immune, reproductive and cardiovascular systems, even at relatively low levels. There is no threshold of exposure under which adverse effes cannot be detected.

While exposure to lead and lead poisoning may be due to many different sources and products – including paints and pigments, electronic waste, cosmetics and toys, traditional medicines, contaminated food and drinking water systems – lead in petrol has been the biggest contributor to global environmental lead contamination.

From the period 1976–1980 to the period 1999–2002, the United States saw a 98 per cent reduction in the proportion of children aged one to five with levels of more than 10 micrograms of lead per decilitre of blood. Other studies worldwide showed a strong correlation between decreased use of lead in petrol and reductions of lead in blood.

Interventions to prevent lead poisoning have demonstrated very large economic benefits. An analysis of the direct medical and indirect societal costs associated with lead poisoning in children in the United States found these to be US$43 billion annually, even at the relatively low levels of exposure to lead at the time. Another economic assessment, in this case for the full lifetime productivity of people, estimated that the increases in child intelligence, and thus their in-lifetime economic productivity, resulting from the removal of lead from petrol produced a benefit of between US$110 billion and US$319 billion in each birth cohort in the United States.

According to Gould, for every US$1 spent to reduce lead hazards, there would be a benefit of US$17–220, a better cost-benefit ratio than that for vaccines, which have long been described as the single most cost-benefcial medical or public health intervention. Another methodology that uses GDP extrapolation applied to published literature showed the global benefits of the phase-out of lead in petrol to be US$1–6 trillion per year, with a best estimate of US$2.45 trillion per year, roughly 4 per cent of global GDP.

Recent evidence on health impacts has led the US Environmental Protection Agency to tighten its rolling three-month average air quality standard for lead from 1.5 micrograms of lead per m^3 of air in 1978 to 0.15 micrograms per m^3 in 2008. The WHO annual ambient air guideline for lead remains at 0.5 micrograms of lead per m^3 of air. The removal of lead from petrol and the consequent reductions in health risks is an outstanding global success story, with the complete global elimination of lead in petrol expected within a few years.

Emerging Issues

The most important new issue in the study of the atmospheric environment is the role of short-lived climate forcers, especially methane, tropospheric ozone and black carbon. A subset of hydrofuorocarbons (HFCs) are also important short-lived climate forcers.

Black carbon particles in the atmosphere have a significant impact not only on human health but also on climate. They darken snow and ice surfaces, lower their albedo and increase their absorption of sunlight, which, along with atmospheric heating, exacerbates the melting of snow and ice around the world, including in the Arctic, the Himalayas and other glaciated and snow-covered regions. This affes the water cycle and may increase risks of fooding. Methane is a powerful greenhouse gas and an important precursor for ozone generation. Methane, black carbon and tropospheric ozone are fundamentally different from the longer-lived greenhouse gases as they remain in the atmosphere for a relatively short time. Reducing black carbon and methane emissions now will slow the rate of climate change within the first half of this century.

A second major emerging issue is the health effect of fne particles of natural origin. Every year, very large amounts of soil-derived dust and particles from wildfres engulf major populated areas. These can include soil particles from arid regions being deposited on coastal cities of China, Saharan dust reaching cities in Africa and the Mediterranean, and dust from drought-affeed inland areas being deposited on cities in the United States and Australia. In addition, smoke from wildfres commonly raises particulate concentrations in Africa, Siberia, the Mediterranean, the United States, South East Asia and Australia. These particles can have major impacts on human health and a recent study suggests that almost 300 000 excess deaths per year can be attributed to fne particles of natural origin. The sources can, however, be at least partially controlled. Major interventions to re-vegetate degraded landscapes are continuing in several countries and the 2003 ASEAN Agreement on Transboundary Haze Pollution is an example of an international agreement aimed at addressing the international transport of particles generated from forest burning.

As understanding of the relationship between particle size, the number of particles and health impacts has improved, concern has grown about the impacts of fne particles (smaller than 2.5 micrometres in diameter) and ultrafne particles (1 micrometre and sub-micrometre sizes) on respiratory and cardiovascular health. As the volume of evidence is rapidly growing, it is likely that in the next few years air quality standards and guidelines to protect health by controlling exposure to ultrafne particles will be developed and become a focus of air quality policy, monitoring and management.

A number of new approaches to address the challenges of climate change have been proposed, including carbon capture and storage and geo-engineering.

Atmospheric Governance and Integrated Approach to Management

It suggests there is no one-size-fts-all solution to most atmospheric problems. The use of targets and timetables that worked for stratospheric ozone depletion may stall negotiations over climate change. Emissions trading schemes that worked well for the reduction of sulphur dioxide in some developed countries may need to be complemented by other measures in developing countries. Many emission sources release both greenhouse gases and air pollutants; some air pollutants have an additional effect on the climate; and reducing consumption of ozone depleting substances also reduces their impact on climate. There will be a growing need for decision-making frameworks and enabling environments that explicitly recognize the integrated nature of the atmosphere.

The elimination of lead from petrol was made easier by a cost-effective alternative that proved easy to communicate to politicians and other stakeholders. With the well-timed support of international initiatives such as the UNEP Partnership for Clean Fuels and Vehicles, one country after another introduced lead-free fuels.

Though there was no global binding agreement on the phase-out of lead in petrol, there are parallels with the phase-out of ozone-depleting substances as a relatively manageable problem with cost-effective solutions and a high level of concern. For the elimination of substances that harm the ozone layer, governments agreed to the Vienna Convention, setting in motion an international negotiation process that culminated in the Montreal Protocol. The protocol became the model for other international agreements that called for a series of targets and timetables to eliminate ozone-depleting substances in the developed world, and for the creation of a multilateral fund to fnance replacement technologies for developing countries beginning to manufacture CFCs. The process of arriving at this agreement helped raise concern, lower costs and clarify complexities.

Progress with other pollutants has been more uneven. For example, in the case of sulphur dioxide, existing technology, afordable abatement costs and growing understanding have made the issue increasingly manageable in much of the developed world. However, while target setting and the installation of fue-gas desulphurization have become common, growth in the number of coal-fred power plants has overwhelmed efforts to reduce emissions. Consequently, levels of acid deposition in East Asia remain high.

The health impacts of particulate matter make its control the highest priority. However, measures can be costly and complex due to a multiplicity of industrial, transport, energy, commercial, domestic and natural sources, especially in developing countries. Various measures, such as technological improvements in vehicles, increased engine efciency, cleaner fuels and particulate flters, have been successfully applied in different cities. Across the developed world urban levels of particulate matter began to fall sharply in the 1950s and 1960s. Cleaner technologies have enjoyed some success in reducing emissions in developing countries, but have not been sustained in rapidly growing cities where high demands for motorization, energy and industrial products have increased aggregate emissions. Both the complexity of the issue and the costs have hampered progress. To reduce exposure to indoor particulate matter national policies, including on rural development and energy, need to be mainstreamed into overall development policy.

Governance issues related to climate change have a high level of complexity, mixed levels of concern and long lead times between actions and benefits, often beyond political timescales. The governance approach for climate change has in many ways followed a similar approach to that of the ozone layer, but with different results due to diferences in the nature of the issues. A growing degree of concern led to a global agreement, the UNFCCC, which allowed negotiation of the Kyoto Protocol. This was intended to initiate the process of reducing anthropogenic greenhouse gas emissions, but even if fully implemented was never designed to be enough to stay within the UNFCCC agreed limit of a 2°C temperature increase.

The approach of developing binding national targets within a global framework has so far not delivered emission reductions that would help achieve climate targets and internationally agreed goals. In the medium term, a promising approach appears to be the development of nationally appropriate mitigation actions – or NAMAs – to encourage countries to contribute to actions in a national context.

The achievement of global climate goals will most likely require a transformational change addressing the main drivers of emissions such as in the way electricity is generated, the efciency with which energy and resources are used, and the management of terrestrial ecosystems. Consumption levels and production processes may require the introduction of approaches such as a circular economy in which material fows are either made up of biological nutrients designed to re-enter the biosphere, or materials designed to circulate without entering the biosphere. Such changes take time and, in the short term, existing cost-effective options need to be rolled out as far and as rapidly as possible to deliver early emission reductions that put the world on a path towards achieving established climate goals.

Mitigating near-term climate change – the warming likely to be experienced over the next two to four decades – is important to prevent damage to vulnerable ecosystems such as the Arctic, and to vulnerable societies such as those in drought- or food-prone regions. Addressing CO_2

is not likely to be sufcient to mitigate warming in these timescales, partly because it is long-lived. Fortunately, near-term warming can be addressed by additional complementary policy action that reduces concentrations of short-lived climate forcers, including black carbon, methane and tropospheric ozone. Addressing these short-lived substances is an example of an integrated approach to atmospheric governance that provides an opportunity for policy development to meet multiple goals in a cost-effective way. The improved awareness of atmospheric brown clouds emphasizes the integration of different atmospheric issues. There is increasing evidence that the Antarctic ozone hole has affeed the surface climate in the southern hemisphere, and further links also exist between climate change and ozone-depleting substances as many are also very strong greenhouse gases. Indeed, avoiding emissions of CFCs has brought about a significant contribution to climate change mitigation.

These interactions and links between different atmospheric issues provide opportunities to improve progress in reaching internationally set goals, maximizing benefits and avoiding policy conficts. To make rapid progress, scientifc knowledge needs to be properly presented to policy makers to allow them to deal more effectively with the complexity of the issues. The way in which options are analysed, costs and benefits weighed up, and evidence-based policies developed, requires considerable improvement. This will require closer interaction between the science and policy communities, increased stakeholder participation, capacity enhancement and technology transfer.

Gaps and Outlook

Concern for the impacts of atmospheric issues at global, regional and national scales has led to considerable efforts to control emissions to meet internationally agreed goals and targets. Some issues have been effectively dealt with. For others there has only been partial success, with improvements in some regions but problems remaining in others.Targets to protect the global atmosphere from air pollution are being met for stratospheric ozone depletion and lead in petrol.

For the majority of the world, however, most air quality guidelines are not being met as there is insufcient implementation of policies. Meanwhile, important ecosystems are experiencing pollution loads in excess of critical thresholds. In the near term, atmospheric issues such as particulate matter and other pollutants could, with adequate commitment and resources, be effectively addressed by the wider implementation of existing policies and technologies.

The current development trajectory, based on existing models of international governance, is unlikely to meet internationally agreed atmospheric goals, especially those for mitigating climate change and reducing the health impacts of pollutants. Carefully selected approaches at national and regional scales need to be encouraged and facilitated by global coordination to increase the chance of reaching the targets in the near term.

Climate change presents the global community with one of the most serious challenges to achieving development goals. Serious impacts from climate change are unlikely to be avoided on the basis of current emission reduction pledges. In the medium term, progress could be made by encouraging further national pledges, taking into account individual country circumstances, and by the wide application of current technological and policy approaches.

Measures to reduce emissions of short-lived climate forcers could contribute to reducing the temperature increase in the near term but, ultimately, a transformation in the way energy is provided and the efciency with which electricity and other resources are used, coupled with a shift in consumption and production patterns and investment in innovation, will be required to achieve long-term climate goals. Such transformative change will also affect other atmospheric problems. But action should start now with the measures that are currently available while transformation takes place. Such action would bring significant benefits, particularly if the atmospheric issues and required policies were considered in an integrated way.

References

Ashmore, M.R. (2005). Assessing the future global impact of ozone on vegetation. *Plant, Cell and Environment* 28, 949964.

Hansen, J., Ruedy, R., Sato, M. and Lo, K. (2010). Global surface temperature change. *Reviews of Geophysics.* 48, RG4004

Levy, M.A., Haas, P.M. and Keohane, R.O. (1993). Improving the effectiveness of international environmental institutions. In *Institutions for the Earth: Sources of Effective International Environmental Protection*. MIT Press, Cambridge, MA.

Stern, N. (2007). *The Economics of Climate Change: The Stern Review.* Cambridge University Press, Cambridge and New York.

3

Principles of Soil Conservation and Management

Changing climate patterns, economic globalization, population growth, increasing use of natural resources and rapid urbanization are putting pressure on terrestrial ecosystems as never before, and virtually all of them are under stress. Biophysical limits on what is available for human use are real and there are strong signals that these limits are close to being reached or have already been exceeded. Even so, the fact that some areas show recent gains in forested area or land reclamation suggests that declines are not inevitable, and indeed that recovery may be possible – even though original ecosystem functions may be modified or pressure on ecosystems may shift elsewhere.

Growing demands for food, feed, fuel, fbre and raw materials create local and distant pressures for land-use change. The cascade of outcomes resulting from these demands is complicated by urbanization and globalization, which separate the production of goods from their consumption over vast distances. The central question is how these demands can be met – or managed – in ways that recognize the joint imperatives of human well-being and environmental sustainability. Addressing this requires careful examination of the social relations and biophysical processes involved in managing terrestrial ecosystems, setting priorities for policies and policy instruments, and considering the likely distribution of implications, both positive and negative.

The fourth *Global Environment Outlook* (*GEO-4*) noted that increased demand for water, waste disposal and food had led to unsustainable patterns of land use and land degradation. It identified forest cover and composition, cropland expansion, intensification of agriculture, desertification and urban development as key topics in land-use change. *GEO-4* concluded that continued inaction on land stewardship, combined with increased climate change, would reduce social resilience, making recovery from future stresses difficult or impossible.

Current State of Agricultural Land

Demands for food and livestock feed are rising rapidly due to population growth, urbanization and changing diets that include more animal products. One of the consequences of these changes is the widespread expansion of agricultural land allocated to livestock, both directly and indirectly through cropland dedicated to animal feed production. At a time when water shortage and land degradation remain threats to food security, accelerated interest in biofuel, feeds and fbre in recent years imposes competing demands on how agricultural land is used.

Agricultural Land and Production Trends

In 2009, there were approximately 3.3 billion hectares of pasture and 1.5 billion hectares of cropland globally, with the extent and proportion of total land area varying greatly across regions. In 2009, all regions except Europe had a greater proportion of land area devoted to pasture than to cropland. Although there has been only a slight increase in total cropland extent over the past decade, there has been a considerable change in the crops grown. Maize is an important crop in all regions other than West Asia, with the area harvested increasing by more than 25 per cent across Africa and Asia and the Pacifc between 2001 and 2010.

In total, approximately 160 million hectares of maize were harvested in 2010. Asia and the Pacifc have the largest area of rice, but Europe and Africa experienced the greatest percentage increases between 2001 and 2010 – about 30 and 20 per cent respectively. The dominant soybean-producing regions are Latin America and the Caribbean and North America, with the United States, Brazil and Argentina the three largest producers. Asia and the Pacifc and Europe are the primary producers of wheat.

Increases in the area used for these crops have been accompanied by overall growth in yields. Globally, the current yields of wheat, maize and rice have been estimated at 64, 50 and 64 per cent of their potential respectively, but the size of the yield gap varies greatly from region to region under the influence of different factors. Larger gaps between actual and potential yields tend to occur where low-input agriculture is practised. Africa and Latin America and the Caribbean – two regions where crop area has expanded since 2001– still have relatively low yields compared to North America and Europe; if region-specifc constraints can be assessed and overcome, there may be potential to increase food production in these regions while minimizing cropland expansion.

Agricultural productivity is limited by biophysical and other factors. Extending conventional agriculture into uncultivated lands requires mechanization to modify the surface, and supplements in the form of fertilizers, herbicides, pesticides and irrigation water. Excessive use of machinery and chemical supplements, howcver, breaks up soil structure, increases erosion, chemically pollutes soil, contaminates groundwater and surface water, changes greenhouse gas fluxes, destroys habitat and builds genetic resistance to chemical supplements. With widespread adoption of intensive, mechanized, high-input agricultural practices, the rate of soil erosion has greatly increased. Erosion in conventional agricultural systems is now over three times higher than in systems practising conservation agriculture, and over 75 times higher than in systems with natural vegetation. Globally, soil erosion is contributing to the decline in agricultural land available per person as degraded land is abandoned. Thus, the yield gains achieved by these methods come with ecological costs.

In continuously cultivated, low-input agricultural systems, rapid declines in soil fertility and yield, together with international commodity price changes, continue to impact human well-being in agricultural communities. Sustainable intensification techniques ofer the potential to improve soil fertility and yields in some situations while avoiding some of the problems of high-input agriculture just presented.

While the future impact of climate change on global food production is difficult to specify, substantial evidence suggests that an increasing number of people will be directly affeed by climate change impacts on agricultural areas.

Consumption Trends

While the proportion of undernourished people has been declining – from 14 per cent of world population in 1995–1997 to 13 per cent in 2010 – the absolute number rose over the same period from 788 million to an estimated 925 million due to population growth. Areas with chronic food insecurity face many obstacles, including regional confícts, weak governance structures and a breakdown of local institutions, all of which affect access to and distribution of food. Many of the world's undernourished people live in areas that are also particularly vulnerable to climate variability. Africa and Asia and the Pacifc were the regions with the lowest average food consumption in 2007, but they were also the regions that had experienced the highest percentage increase. While the Asia and Pacifc region is home to the largest number of undernourished people, at 578 million, sub-Saharan Africa has the highest proportion of undernourished people – about 30 per cent of its population in 2010.

Forests

Forests play a crucial role in terrestrial ecosystems and provide a multitude of services such as shelter, habitats, fuel, food, fodder, fbre, timber, medicines, security and employment; regulating freshwater supplies; storing carbon and cycling nutrients; and helping to stabilize the global climate. Historically, forests have been under pressure due to increasing demands for shelter, agricultural land, meat production, and fuel and timber extraction, but in recent decades this pressure has increased due to competing demands for agricultural expansion and biofuel production, rapid urbanization and infrastructure development, and increased global demand for forest products. Forests are also under increasing stress from changes in mean annual temperatures, altered precipitation patterns, and more frequent and extreme weather events.

Forest Area

Forests cover just over 4 billion hectares, 31 per cent of the world's total land area. The majority of these are boreal forests extending across northern and central Russia and much of Canada and Alaska. Large expanses of tropical forest are found in the Amazon, Africa's Congo Basin and parts of South East Asia. Temperate forests remain in a patchy distribution across the United States, Europe and the Asian mid-latitudes.

The rate of forest loss from both deforestation and natural causes is slowing, but remains alarmingly high. At the global level, annual forest loss decreased from 16 million hectares in the 1990s to approximately 13 million hectares between 2000 and 2010. The highest rates of tropical forest loss over this period occurred in South America and Africa. Some rapidly developing countries that sufered extensive deforestation in the 1990s, including Brazil and Indonesia, have significantly reduced their rates of tropical forest loss, while less developed nations in Latin America and Africa continue to experience high rates of loss. Although much of the developed world has experienced net reforestation since the late 1800s as a result of rural-urban migration and farm abandonment, natural factors such as drought, forest fire and insect attacks have exacerbated forest loss in recent decades. However, the key drivers of forest loss are population growth, poverty, economic growth, land pricing, international demand for timber and other forest products, insecurity of the rights of local people, and incomplete valuation of forest ecosystems.

Forest Plantations

Forest plantations, generally cultivated for industrial purposes, increased by 50 million hectares globally between 2000 and 2010, reaching 264 million hectares or 7 per cent of the total forest area. Asia accounted for 28 million hectares, or 58 per cent of this increase. Generally, monoculture plantations tend not to enrich local biodiversity, but they do provide ecosystem services including timber, carbon and water storage and soil stabilization.

Productive and Protective Forest Area

The global forest area designated for the production of timber and non-timber products declined from about 1.16 billion hectares in 2000 to about 1.13 billion hectares in 2010, an annual decrease of about 2.91 million hectares or 0.25 per cent. However, the global forest area designated for protection of soil and water increased from about 272 million hectares in 2000 to about 299 million hectares in 2010, an annual increase of some 2.77 million hectares or 0.97 per cent. Similarly, the global forest area designated for biodiversity conservation has increased from around 303 million hectares to about 366 million hectares, an annual increase of about 6.33 million hectares or 1.92 per cent. The main reason for the decrease in forest area designated for production is deforestation, and for the increase in protective forest area is aforestation.

Forest Management and Certification

The Forest Stewardship Council (FSC) and the Programme for the Endorsement of Forest Certification (PEFC) are the two main forest management certification organizations. There was an increase in certified forests of about 20 per cent per year between 2002 and 2010 under these two agencies. However, in 2010 about 10 per cent of the total forest area was under FSC- or PEFC-certified forest management. These trends indicate that while there is improvement in forest management, much work remains to be done.

Forest Carbon Stock

Forests are considered an important sink for atmospheric carbon dioxide (CO_2) because of their ability to store carbon in their biomass and soil. More than 75 per cent of the total terrestrial biomass carbon stock and more than 40 per cent of the soil organic carbon stock are found in forest ecosystems. In the 1990s, forest carbon sequestration was equivalent to approximately one-third of carbon emissions from fossil fuel combustion and land-use change. Boreal forests store more carbon in their soils than tropical forests, while tropical forests store much more carbon in their plant biomass. Pan *et al.* estimate that global forest systems constituted a total carbon sink of 2.4±0.4 billion tonnes of carbon per year from 1990 to 2007.

Fires are a major source of greenhouse gas emissions from forests. Boreal forest ecosystems are prone to frequent and severe wildfres leading to large carbon emissions. Amiro *et al.* estimated that during the period 1949–1999, on average 2 million hectares of the Canadian boreal forest burnt annually (ranging from 0.3 to 7.5 million hectares in any given year), emitting a yearly average of 27±6 million tonnes of carbon (ranging from 3 to 115 million tonnes in any given year). Sukhinin *et al.* estimated that on average 7.7 million hectares of area burnt annually between 1995 and 2002 in eastern Russia and that 55 per cent, 4.2 million hectares, of that area was forest. Gillett *et al.* found that recent increases in area burnt in Canada are a result of anthropogenic climate change. In the future more fres, more area burnt and longer fire seasons may be expected in temperate and boreal regions.

Drylands, Grasslands and Savannahs

Drylands, grasslands and savannahs experience high spatial and temporal variability in rainfall, resulting in dramatic diferences in plant growth, habitats and human livelihoods. Drylands cover approximately 40 per cent of the world's land surface and are home to more than 2 billion people, 90 per cent of whom are in developing countries. However, the spatial extent of drylands remains uncertain due to variations in ecosystem sub-types, data variability and the different classes and thresholds applied to remotely sensed data, making global comparisons challenging. Grasslands range from very dry, almost desert-like, to humid types. Savannahs are mixed tree-grass ecosystems, ranging from almost treeless grasslands to closed-canopy woodlands that occupy large areas in the tropics and sub-tropics, particularly in Africa, Latin America and Australia.

Trends in Drylands, Grasslands and Savannahs

Fluctuations in precipitation are a major driver of change in plant cover, but grazing intensity has also been directly linked to long-term dryland degradation. Transformation of rangelands to cultivated croplands is leading to a significant, persistent decrease in overall dryland plant productivity. Sietz *et al.* indicated that the most important factors causing vulnerability in drylands are water stress, poverty, soil degradation, natural agronomic constraints and isolation from political centres.

Net primary productivity (NPP) is the net amount of carbon captured by vegetation through photosynthesis each year. Approximately 2 per cent of global terrestrial NPP is lost yearly due to dryland degradation, equivalent to 4–10 per cent of dryland potential NPP. Sustainable development in drylands will rely on techniques that improve soil fertility, conserve soil and water and increase agricultural efciency, such as mulch farming, conservation tillage and diverse cropping systems.

As an international response to desertification, land degradation and drought in drylands, the United Nations Convention to Combat Desertification (UNCCD) was adopted in 1995 and has since been signed by 194 Parties – 193 countries plus the European Union. Following mixed results in its initial implementation phase, Parties to the Convention adopted a ten-year strategic plan for 2008–2018 to revitalize it. The plan includes a results-based management approach built on a set of specifc objectives and indicators, and a new monitoring, assessment and reporting process – the performance review and assessment of implementation system.

WETLANDS

In 2003, the European Space Agency, in collaboration with the secretariat of the Convention on Wetlands of International Importance (Ramsar Convention), launched the GlobWetland project to demonstrate the current capabilities of Earth observation technology to support inventorying, monitoring and assessment of wetland ecosystems. The project revealed a major gap between the findings of the Earth observation and wetland communities, with considerable inconsistency in global wetlands estimations.

The conversion of wetlands continues. For both inland and coastal wetlands, the most salient drivers of change are population growth and increasing economic development, which in turn promote infrastructure development and land conversion including agricultural expansion. Other direct drivers affeing wetlands are deforestation, increased withdrawal of freshwater, diversion of freshwater fows, disruption and fragmentation of the landscape, nitrogen loading, overharvesting, siltation, changes in water temperatures and invasion by alien species. In 14 deltas analysed by Coleman *et al.*, over half of the studied wetland area of 1.6 million hectares had been irretrievably lost over a 14-year period due to natural causes, conversion to agriculture or industrial use. Global climate change may exacerbate the loss and degradation of coastal wetlands. For example, Syvitski *et al.* analysed the effes of human activities on delta subsidence, susceptibility to fooding and vulnerability to sea level rise, concluding that the area of deltas at risk of fooding could increase by more than 50 per cent by the end of this century.

The deforestation, drainage and conversion to agriculture of peatland results in substantial emissions of CO_2 and nitrous oxide. Globally, peatlands cover 3 per cent of the world's land surface, about 400 million hectares, of which 50 million hectares are being drained and degraded, producing the equivalent of 6 per cent of all global CO_2 emissions. Avoiding further wetland degradation could result in significant climate change mitigation.

Because of increasing demand for land for food, feed, biofuels and materials, the loss of wetlands and associated ecosystem services is likely to continue. Globally, coastal wetlands such as mangroves are continuing to decline by more than 100 000 hectares, over 0.7 per cent, per year, but that rate of loss has slowed relative to the 1 per cent per year of the 1980s. Although, in most regions, rates of loss have decreased compared to the 1980s and 1990s, mangrove losses in Asia accelerated again during 2000–

2005. Despite these losses, the Asia and Pacifc region holds the largest spatial extent of mangrove systems – more than 50 per cent of the global total. Other major mangrove areas are in northern Latin America, Eastern and Western Africa, and the Red Sea.

Polar Regions

The Arctic's permafrost – the top 3.5 metres of soil that remains permanently frozen for 24 months or more – contains the largest deposits of organic carbon on Earth. But due to some of the most rapid warming on the planet, with temperatures in the permafrost having already risen by up to 2°C over the past two to three decades, this is likely to become a substantial source of carbon emissions over the next century. The Arctic's tundra and boreal forest ecosystems currently act as a carbon sink, but it is possible that the Arctic region will become a net emitter over the course of the 21st century as up to 90 per cent of near-surface permafrost is expected to disappear due to thawing by 2100.

Methane emissions, primarily from wetlands, also play an important role in the carbon balance of the Arctic. Although just 2 per cent of global methane emissions originate in the Arctic, this region has seen the largest proportional increase in emissions, rising by nearly a third between 2003 and 2007. Some of these emissions originate in the escape of methane from sequestration within hydrate crystals frozen beneath permafrost. These methyl hydrates also occur in abundance beneath the deep ocean foor and within continental shelves. Methane is 25 times more effective at trapping heat in the atmosphere than CO_2 over a 100-year horizon.

Other climate-related land changes occurring in the Arctic include the northward movement of tree lines, woody vegetation encroachment into the tundra, and a longer growing season – resulting in an increase in plant productivity. Whilst these processes remove CO_2 from the atmosphere, it is likely that the release of carbon from thawing permafrost and other processes will outpace CO_2 sequestration by vegetation.

Environmental changes such as the northward advance of tree lines, combined with rapid industrial development, create challenges for traditional livelihoods such as reindeer herding.

Access to many areas on land, especially in northern Canada and Russia, is becoming more difficult as ice roads melt earlier and freeze later, severely affeing communities and industrial development. At the same time, because the seasonal Arctic Ocean ice cover is decreasing in area, volume and duration, new economic opportunities are presenting themselves, including increased tourism, forestry, agriculture, and expanding oil, gas and mining developments. Nonetheless, some communities in the Arctic most affeed by thawing permafrost and/or coastal erosion are being forced to relocate, and further research is needed to foresee how living conditions are likely to change and to evaluate possible adaptation options, taking the region's indigenous peoples into particular consideration.

At the southern pole, the landmass of Antarctica also has a profound effect on the Earth's climate and ocean systems. However, in contrast to the Arctic, the Antarctic land mass is 99 per cent covered by glacial ice.

Urban Areas and Human Infrastructure

Urbanization has progressed at an extraordinary rate in recent decades and this growth is projected to continue throughout the century. Urban areas are the hubs of social processes, driving many changes through material demands that affect land use and cover, biodiversity and water resources, locally to globally. Nevertheless, if well planned, urban areas can reduce the overall pressure on land resources of a growing population.

Satellite-based studies calculate urban land cover at less than 1 per cent of the planet's total land surface. However, the impact of urban areas on the global environment cannot be measured only by their physical expansion. Some studies estimate that 60–70 per cent of total anthropogenic greenhouse gas emissions are directly or indirectly related to urban areas, with a few wealthy cities contributing the majority of emissions. It is the concentration of population, economic activities and wealth generation in urban areas that drives their impact on the global environment, with demands for food, energy, water and production materials that have significant consequences for land-use change around the world.

Most of the understanding of urbanization as a land-change process is based on individual case studies that reveal significant diferences in urbanization processes between regions and countries, and even within countries. Ecological footprint analyses of cities provide a symbolic parameter illustrating the impacts of those diferences on the local and global environment. For example, the inhabitants of a typical city of 650 000 inhabitants in the United States collectively require 3 million hectares of land to meet their domestic needs, while those of a similar-sized city in India require just 280 000 hectares.

Urban Trends

The UN Population Division projects that between 2007 and 2050, the world's urban population will increase by more than 3 billion, with almost all future population growth expected to take place in the cities and towns of developing countries. By 2050, more than 70 per cent of China's population and 50 per cent of India's is likely to be urban, with China expected to have 30 additional cities of more than 1 million inhabitants and India 26.

Urbanization is not a homogeneous process. Recent studies suggest a significant increase in land requirements for urban uses in the next 40 years – potentially an additional 100–200 million hectares. This increase is expected to occur primarily in sprawled patterns and to have major effes on greenhouse gas emissions, air pollution and waste management.

Very large cities exert local and global impacts on the environment, for example the emission of greenhouse gases or aerosols that have a dimming effect in the atmosphere. Small and medium cities, despite their own environmental impacts, may have better opportunities to improve their relationship with the environment and social well-being, particularly in low-income and middle-income countries, where population will concentrate in the future. Only 12 per cent of the total urban population in developing countries lives in very large urban areas of more than 10 million people, while 40 per cent lives in cities of less than 1 million.

Major Issues in Land Change

The changes in land use presented in this are a product of complex interactions between human actions and biophysical processes. International goals provide one set of guidelines for land management, but these are often overshadowed by other pressures and competing demands. Here, four major themes are explored that help to explain the apparent movement away from achieving land-related goals:

- economic growth at the expense of natural capital;
- competing demands for land;
- increased separation of production from consumption; and
- governance challenges related to sustainable land management.

Each is illustrated with examples of impacts on land resulting from these pressures, as well as opportunities to move land management decisions towards social and ecological outcomes in line with international goals.

Economic Growth and Natural Capital

The global economic system is based on the pursuit of perpetual and unsustainable growth. Distorted incentives have reduced natural capital, while often rendering attempts to curtail resource or energy use politically problematic. Simply put, economic growth has come at the expense of natural capital.

Today, many terrestrial ecosystems exhibit degradation and reduced resilience. This can be linked to the failure to account for the vital functions of these ecosystems in economic cost-benefit analyses. For example, financial pressures have encouraged the irrigation and subsequent salinization of vast dryland areas, making them very difficult to rehabilitate. Wetlands continue to be drained for agriculture and urban development, destroying their ability to regulate water quantity and quality and bufer against extreme weather events. Deforestation and forest degradation produce financially attractive short-term returns, but global natural capital losses have been recently estimated at between US$2 trillion and US$ 4.5 trillion each year.

Ecosystems have priceless spiritual, aesthetic and cultural dimensions. They are also the cornerstones of economies, but their real value remains effectively invisible in national profit and loss accounts. Allowing the privatization of profits from the extraction of natural capital at the expense of more innovative and equitable land management approaches is a pervasive problem across all land covers and uses. Incentives that are narrowly focused on economic growth often encourage land management that degrades ecosystem services, while including and valuing ecosystem services in accounting systems can help protect and enhance them. Successful strategies rest on improving understanding of ecosystem functions and building that understanding into policies and institutions. Indeed, recognition of the multiple uses and multiple values of ecosystems can be used to leverage resources for their protection.

Over the past two decades, payment for ecosystem services (PES) has gained attention as a mechanism with the potential to account for services provided by ecosystems in market transactions, build bridges and balance interests between the users and providers of these services, and deal with the linked challenges of conservation and poverty alleviation. Payment for ecosystem services involves a suite of approaches linked to a broad central idea: "the transfer of resources among social actors with the objective of creating incentives to align individual and/ or collective land-use decisions with the social interest in the management of natural resources".

The concept of PES ofers several advantages over conventional conservation approaches: it complements command-and-control and polluter-pays principles with more flexible, incentive-based approaches; it is conditional and voluntary, with the potential to promote equity, accountability and cost effectiveness; and it can produce co-benefits for livelihoods and contribute to poverty alleviation. Positive land-use outcomes have been achieved through some PES initiatives in, for example, Colombia, Costa Rica and Nicaragua, where tree cover has increased and degraded pasture decreased due to a regionally integrated PES project.

However, groups who oppose the idea of nature being commoditized or traded have criticized the concept. Furthermore, despite promising initial benefits such as increased land-tenure security, current evidence of PES's cost effectiveness and the conditions under which it has positive environmental and socio-economic impacts remains inconclusive, particularly in developing countries with weak governance.

Challenges ahead for PES focus on cost effectiveness, monitoring capacity, enforcement, transparency and accountability, and clear boundaries to land access and tenure rights. Taking into account social norms and culture, building trust between actors and dealing with power relations will ultimately define benefit allocation strategies and successful long-term implementation of PES.

Competing Demands for Land

The challenge of feeding a growing human population has been compounded by rising afuence in some regions. Changing diets and increasing demand for biofuels and other industrial materials such as timber have intensified competition for land and pressures on terrestrial ecosystems.

Food Security

To meet the Millennium Development Goal (MDG) 1c on reducing hunger, global food production will have to increase and food distribution improve.

To meet MDG 7 and other environmental goals, agriculture needs to reduce its current environmental impacts.

Although estimates vary, the Food and Agriculture Organization of the United Nations (FAO) projects that to reduce the proportion of developing countries' populations that are chronically undernourished to 4 per cent in the year 2050, world food production will need to increase by 70 per cent from 2005 levels. Although food consumption per person is increasing across all regions, it is unevenly distributed and the number of malnourished people continues to rise as more grain is diverted to produce meat for those who can aford it. Livestock and poultry can serve as an important source of protein in areas of chronic food insecurity and provide an important bufer in times of crop failure, but a disproportionate share of agricultural land is dedicated to meat and dairy production for consumption in developed countries. Such land use is less efcient in meeting global food needs and comes with greater environmental consequences than cropland. For example, it is estimated that the amount of grain fied to livestock in the United States is more than seven times that consumed directly by the population.

Meanwhile, about one-third of all food that is produced for human consumption is wasted or lost – approximately 1 300 million tonnes annually. The concept of food security moves beyond the question of whether adequate food is available and considers whether people have physical and economic access to food. This draws attention to a broad set of social and political issues related to food distribution.

It will be challenging to meet future global demand for food while avoiding, or at least mitigating, negative impacts on forests, wetlands and other ecosystems – and at the same time reducing poverty, supporting livelihoods, and ensuring food safety and animal welfare. There is little debate that more land will have to be allocated to agriculture, but this will not be sufcient without increasing yields and reducing losses along the food supply chain. Climate change is likely to complicate matters further by affeing crop yields in many areas.

A variety of agricultural approaches is likely to provide the best outcomes for food security and environmental well-being. High-input, intensive agricultural methods undeniably increase agricultural yields, though these gains may come at the expense of long-term soil fertility. Location-specifc approaches are also needed in order to achieve sustainable land use based on biophysical as well as socio-economic considerations, while agroecology and urban agriculture can contribute to the global food supply. Agricultural practices that conserve soils and nutrients such as no-till farming can complement efforts to restore degraded and abandoned agricultural land.

Meeting the global need for food will be one of the most important challenges of this century, and a portfolio of solutions including conservation agriculture, high-yielding cultivars, and efcient and carefully managed use of fertilizers is needed rather than promotion of a single strategy. Advocates of genetically modified crops point out their potential to increase yields while reducing the use of agricultural chemicals, although resistance to their use remains, in part, due to the uncertainty of potential risks to human health and further loss of agricultural biodiversity.

Meat Production

Meat production has increased significantly during the past two decades, outpacing the rate of population growth over the same period. Large diferences in meat consumption exist both within and between countries, ranging from an average of 83 kg per person per year in North America and Europe to 11 kg per person per year in Africa. Population growth, urbanization and increasing incomes are expected to continue to raise demand for meat, particularly in developing countries.

The environmental impacts of meat production depend on intensity, extent and management. Nonetheless, growing demand for meat worldwide has been an important driver of deforestation in South America, as forest is cleared to plant soy for livestock feed. As meat production has grown, so has the area harvested for soybean crops, which expanded to 98.8 million hectares in 2009 from 74.3 million hectares in 2000, and 50.4 million hectares 30 years ago.

An increasing demand for meat has the potential to compound rangeland degradation. Livestock production accounts for over 8 per cent of global freshwater use and is among the largest sources of water pollution leading to eutrophication, algal blooms, coral reef degradation, human health issues, antibiotic resistance and disruption of nutrient cycling. Considering the entire commodity chain, including deforestation for grazing and forage production, meat production accounts for 18–25 per cent of the world's greenhouse gas emissions, which is more than global transport. Reducing meat consumption in regions where it is relatively high could thus bring a range of environmental benefits.

Biofuels

An urgent search for renewable sources of energy has resulted in policies that promote the use of biofuels. Increased production of crops that can be used for multiple purposes including food, feed or fuel – such as oil palm, soy, maize and sugar cane – is indicative of this trend. However, subsidies that promote biofuels have been linked to distortions in the world food system, leading to increases in food prices. Recent changes in the linked production of food, feed and fuel have far-reaching impacts for ecology as well as for social relations and vulnerability. While no energy source is completely problem-free, biofuels present particular challenges to land use and terrestrial ecosystems. This, combined with the recent rapid increase in their production, is the reason for examining them here.

While a major motivation for promoting and investing in biofuels has been the desire to reduce greenhouse gas emissions, recent research shows that their emissions balance varies widely depending on which crops are grown, where, and which production methods are used. Biofuel crops have been linked to deforestation, for example in Indonesia, and to encroachment into conservation lands. Once these land-use changes are taken into account, the biofuel carbon balance can become negative, meaning that more carbon is released producing and using biofuels than the equivalent amount of energy from fossil fuels.

Crop-use changes stemming from demand for biofuels have already been observed. For example, in 2007, the United States converted 24 per cent of its corn to ethanol, supported by government subsidies. The US Renewable Fuels Standard of 2007 mandated an increase in biofuel production from around 6.5 billion litres (1.7 billion US gallons) per year in 2001 to 136 billion litres (36 billion US gallons) per year by 2022. Also in 2007, US farmers planted the largest area in maize since 1944: 37.8 million hectares, an area 20 per cent bigger than in 2006. This crop change, which was subsidized, resulted in calling back into production many set-aside lands in the Conservation Reserve Program (CRP) that used to help check surpluses, maintain price levels and promote an ecological balance. Between late 2007 and March 2009, the total area of CRP land in the United States dropped from 14.9 million to 13.6 million hectares. In other words, close to 1.3 million hectares of conservation lands were lost in just over a year.

A similar trend can be seen in the European Union (EU), particularly Germany, whose production capacity for biodiesel increased fivefold between 2004 and 2008. Although Germany's rapeseed cultivation reached 1.53 million hectares in 2007, a little over half of which was used for fuel to meet its EU mandatory biodiesel blending target, Germany needs an additional 1.8 million hectares of rapeseed, which can be done only by increasing the conversion of permanent grassland – similar to the US CRP. However, Germany has already used its maximum allowable 5 per cent of grassland under the EU Common Agricultural Policy. Such constraints on agricultural expansion in the United States and European Union help explain the push to outsource biofuel (and food) production to other countries.

Critiques of biofuels have been accompanied by the emergence of alternatives. For example, under certain conditions, community-based biofuel production for local consumption can be desirable, such as in Brazil where some small-scale farmers produce fuel for their own vehicles and equipment. To be considered benefcial, biofuel production should satisfy multiple criteria, including real energy gains, greenhouse gas reductions, preservation of biodiversity and maintenance of food security. Indeed, the principles of ecoagriculture can be applied to help guide biofuel production towards the mutual attainment of production, livelihood and conservation objectives. While such systems represent only a tiny portion of overall biofuel production, they provide an opportunity for equitably distributed alternative fuels to benefit land-based ecosystems, for example by reducing charcoal production.

Timber and Wood Products

Forests are the main source of timber for fuel, industry, pulp, paper and wood-based composites. Key factors contributing to the rise in consumption are population and economic growth. In addition, an increase in the absolute number of people living in poverty, especially in rural areas, and continued urbanization contribute to the growth in consumption of wood fuels, while enhanced economic growth in emerging economies contributes to the increase in consumption of paper and paper products.

Protected Areas

Protected areas are an important mechanism for the conservation of vulnerable environmental resources, although there are controversies as to whether they sometimes come at the expense of the livelihoods of local people. Rates of deforestation are much lower within reserves than outside them, and some research cites the positive benefits that protected areas have on the conservation of ecosystem services.

But when the underlying pressures imposed by local populations are not adequately considered, substantial monitoring and enforcement are needed to enforce rules designed to sustain natural resources, and governance has been found to be most effective when local users participate in the design and implementation of natural resource governance.There is also some evidence of spill-over effes in countries that enact conservation policies, for example by increasing cereal imports from elsewhere. Protection in a given area has also been found to contribute to deforestation on the adjacent land to which displaced human populations have moved.

Despite the growing area of land with protected status – currently almost 13 per cent of the planet's terrestrial area is under some degree of protection – policy makers should not rely solely on this mechanism to preserve natural resources. Instead, they should develop capacity for adaptive management strategies that produce the best institutional ft for natural resource problems while taking into consideration the need to protect local property rights and local livelihoods.

Separation of Consumption from the Impacts of Production

Urbanization and globalization contribute to the separation between places where resources and goods originate and where products are consumed. Recent research suggests that the spatial distance between production and consumption is both significant and growing. As a result, many of the ecological costs of consumption are borne by people and places increasingly far from consumption sites. While urbanization draws people into densely populated spaces and concentrates demand for food, materials and consumer products, globalization and trade facilitate the movement of people and goods, making both regional and international transfers of resources and fnished products possible. Large-scale land acquisitions to supply food, fodder and other forest products as well as other natural resources to markets in distant countries are both a recent outcome of and a contributor to the separation of production and consumption. If carefully planned and managed, urbanization and globalization can present opportunities to increase efciency of resource use.

Drivers of Increased Separation

Urbanization affes land use and land cover, water use and biodiversity at local and regional scales through social processes that drive consumption patterns and material demands. Higher purchasing power among many urban workers contributes to improved quality of life, but at the cost of new challenges for natural resources and environmental management. For example, Western-style diets are increasingly being adopted globally in urban areas. Similarly, improved urban lifestyles are accompanied by higher consumption of water and energy and increased carbon emissions. These urban consumption patterns intensify stresses on distant as well as local ecosystems.

Globalization is not new, but its current iteration has some distinct features. Lower trade barriers, improved communication technologies and relatively cheap transport have all encouraged countries to become increasingly specialized in their economic activities and reliant on international trade to connect products and scrvices with distant markets. While international trade can make use of strategic advantages to produce goods in an efcient way, it also makes it easier to externalize both environmental and social costs. The well-being of individuals in one place is often based on the degradation of the environment elsewhere, for example by non-renewable resource extraction. Meanwhile, both resources and pollution are embedded in trade, and countries that place greater emphasis on free-market economic policies have been linked to higher levels of environmental degradation. The challenge for the global economy is to encourage the best of what it can ofer in terms of efcient resource use while taking measures to reduce the occurrence, concentration and transfer of environmental and social costs.

Land Deals

Recent changes in production patterns can be linked to the convergence of food, energy, environmental and financial crises, and a continuing surge in the mineral and timber industries. These interactions have brought corporations and some national governments, based in the global North and South, to forge widespread land deals, sometimes referred to as land grabs, in distant countries. The UN Committee on Food Security suggests that such large-scale land acquisition now involves close to 100 million hectares. Concentrated in the global South, these land deals are intended to produce food, feed, biofuels, timber and minerals, usually for export. This ongoing global rush for land is altering land-use patterns and social relations, and involves a new combination of people and pressures. Given the rapid pace of recent developments and projected growth in demands for food, feed, biofuels and materials, it is likely to have major impacts on future land use.

The 2007–2008 food price spike inspired multi-sectoral investors to purchase or lease land for food production and export. At the same time, biofuel blending requirements in the EU and many other countries have provided another impetus for external land deals and land-use change. This has directly and indirectly inspired the expansion of oil palm plantations in Colombia, Guatemala, Indonesia and Malaysia, sugar cane ethanol production in Brazil and Southern Africa, soy cultivation in Argentina and Brazil, and the planting of jatropha in Ghana and India, amongst other developments. The emerging pattern of production in these newly opened sites is large-scale, industrial monoculture. Even in cases where contract growing with smallholders is promoted as a key component of new enterprises, monoculture and industrial production methods are adopted, for example in the oil palm sector in Indonesia.

In theory the term marginal lands, often applied to land deals, refers to lands that are far from road networks, are not irrigated, and are not used for intensive commercial agriculture. However, in practice there are indications that land deals have encroached on prime agricultural lands, suggesting that investors do not want to invest in lands with little access to water sources or transport infrastructure.

Displacement of local, including indigenous, people is a potential outcome of these land deals. This becomes a problem if people have nowhere go to seek employment or construct livelihoods. This has happened in several sites of current land deals, pushing people to further crowd urban spaces or into more fragile environments such as remaining forest, higher slopes or river banks. For example, in the Democratic Republic of the Congo, large-scale agricultural investment has reportedly pushed local farmers into a national park. But not all land deals have led, or will lead, to dispossession. Diferent outcomes of land deals for the rural poor are illustrated by McCarthy in Jambi, Indonesia, where three villages showed three broad trajectories: dispossession, relatively successful incorporation into the oil palm enclave, and adverse incorporation with precarious employment and livelihoods.

There are competing views on how to respond. One position argues that land deals ofer both opportunities and threats, and that opportunities can be harnessed and threats managed by promoting a voluntary land-deal code of conduct. In contrast, proponents of minimum human rights principles argue that voluntary codes may be insufcient to ensure that agricultural investment "benefits the poor in the South, rather than leading to a transfer of resources to the rich in the North". An in-between position is reffeed in the Voluntary Guidelines for the Democratic Governance of Natural Resources promoted by FAO, which, unlike corporate-led codes of conduct, bind member states to mandatory reporting. How these viewpoints unfold remains to be seen.

Land Governance

Many of the challenges for sustainable land management stem from underlying weaknesses in land governance systems. Generally, there are three components of a governance system: actors and organizations, institutions, and practices. Incompatibility between these is one of the most common reasons for the lack of successful transition from resource-extractive to sustainable management of land resources. For example, various countries have redirected their policies and management rules towards sustainable forest management, but due to structural and cultural resistance in forestry organizations, management practices have not changed to the expected level. Other common features of poor land governance are low levels of transparency, accountability and participation in decision making, and a lack of capacity amongst the actors and organizations responsible for land management.

Land governance includes structures ranging from totally centralized to completely decentralized. A major challenge is to find the best governance system, which depends on existing governance alongside the social, economic and environmental conditions and their dynamics.

Market-based Approaches

Heightened interest in carbon sequestration has inspired new incentives and fnancing for ecosystem protection. Local and global initiatives have started to invest in market-based climate approaches that attach a financial value to the carbon stored in forests, ofering incentives for developing countries to invest in low-carbon development. One such opportunity – Reducing

Emissions from Deforestation and Forest Degradation (REDD) in developing countries – has emerged as an important component of a global strategy to reduce emissions while generating financial fows from North to South. Since its inception, REDD has evolved into REDD+, which now goes beyond deforestation and forest degradation to include conservation, sustainable forest management and enhancement of forest carbon stocks. Evidence of the potential for carbon sequestration in drylands and grasslands is accumulating in support of REDD+ programmes for these ecosystems as well as forests.

At this stage, REDD+ has not been incorporated into any formal international carbon market, but it is likely to form a key element of a post-Kyoto climate change treaty by promoting the avoidance of deforestation and allied measures as eligible activities for countries seeking to meet their obligations. Carbon offset payments would encourage developing countries to reduce national deforestation rates, while REDD+ could include incentives to promote aforestation, reforestation and improved forest management. Research suggests that when appropriate techniques are used, forest restoration is a cost-effective means of sequestering carbon while providing abundant social and ecological benefits.

Proponents from both science and policy believe that REDD+ will not just conserve forests; they also consider it one of the most cost-effective carbon abatement options worldwide. With the right safeguards in place, REDD+ could ofer crucial new incentives for achieving sustainable development goals – which have proved elusive since the 1992 Rio Earth Summit – by simultaneously enabling biodiversity conservation, watershed protection, capacity building in tropical forest nations and poverty alleviation for rural communities.

Much of the debate around REDD+ has focused on its international aspects. However, its success will largely depend on allocating benefits at the local to national levels and creating domestic safeguards to prevent perverse incentives and the marginalization of forest-dependent communities. To this end, some stakeholders are concerned that REDD+ could pose new risks to already vulnerable populations through restricted access to land, tenure insecurity, confict over resources, centralization of power, and distortion effes in local economic systems. These observers caution that REDD+ will only achieve lasting results if it is suitable for adaptation to the particular circumstances of relevant countries and can meet the needs of local people while building their capacity.

The risks and opportunities for REDD+ will depend on several factors, including how it will be fnanced and implemented. Many challenges are shared by forest countries, but responses and solutions will often have to be developed according to country-specifc and local characteristics. Ultimately, if REDD+ is to be successful, it must generate substantial revenues to implement conservation and sustainable forest management while supporting rural poverty reduction and livelihoods. At the same time, it must recognize the dynamic complexity of global systems, where cause and effect are often distant in time and space.

Land Management and Decentralization

Governance plays a major role in how land resources are monitored and used and how environmental protection is enforced. Proponents of decentralized natural resource management suggest that giving local-level ofcials greater responsibilities should result in more efcient, flexible, equitable, accountable and participatory governance. Local-level decision makers often know more about local conditions and are therefore well positioned to develop new management solutions. This is important from the perspective of adaptive management and providing decision makers with the flexibility to quickly develop solutions to unforeseen problems. But decentralization is only effective if local governments have the financial resources and technical capacity to monitor environmental change. Positive outcomes from decentralized environmental governance are also unlikely in the absence of public participation in local government decision making; this emphasizes the importance of developing the capacity of local-level stakeholders in the sustainable management of land systems.

Capacity Building for Sustainable Land Management

Capacity building recognizes the knowledge systems, perspectives and values of all stakeholders and uses an in-depth understanding of how a resource system functions. As sustainable land management requires a different set of organizational, technical, economic, environmental and managerial skills from that of many land managers, building the capacity of all actors and organizations can be central to its successful integration.

Land degradation in dryland ecosystems provides an example where the lack of capacity - scientifc, technical and collaborative - limits success in addressing environmental problems. Degradation in dryland systems is driven by multiple causes and characterized by complex feedbacks that are made worse by global climate change. Despite concerted efforts and a wide array of initiatives, drylands continue to be threatened because of lack of agreement on the underlying driving mechanisms, characteristics and consequences of degradation. Long-term harmonized data are necessary not only to understand the root causes of observed changes, but also to forecast and disentangle those, possibly irrevocable, impacts of global change from the often more temporary or local variability induced by other human activity. These data gaps, and the subsequent lack of capacity and common strategies among dryland nations, can severely hamper progress towards internationally agreed goals on dryland conservation and rehabilitation.

Outlook

Complex forces are affeing land resources, some at dramatic rates of change and with diverse regional and national characteristics. Certainly some land conversion trends are on an unsustainable trajectory, as global population growth and rising consumption exert ever greater pressures on land. Continued deforestation, wetland conversion and dryland degradation are of particular concern. An increasing portion of the pressure on tropical forests is shifting from the activities of small household farms to large industrial plantations producing soy, meat and dairy products, palm oil, sugar cane and other products destined for global markets. Land degradation continues to hamper soil productivity and ecological functions in many regions. At the same time, there is significant potential to reduce greenhouse gas emissions from agricultural production. Two phenomena that have arisen since *GEO-4* are the expansion of biofuel production and a growing number of land deals in developing countries. These and other processes are unfolding rapidly. While their longer-term implications remain uncertain, early evidence of their social and environmental consequences should be closely considered. In combination, these processes are seriously affeing the environment in several regions and require urgent attention.

Data and Monitoring Gaps

One key to avoiding environmental damage is to effectively monitor environmental trends, yet major data gaps limit the ability to avert unwanted outcomes. Global data on land degradation have not been updated for a long time, although new estimates using satellite material are being developed. Datasets exist for land cover but do not always adequately represent areas that have experienced selective cutting or other types of modification. Forest cover losses in boreal and temperate forests are not as well studied as those in tropical forests, while evidence is still emerging of the significant carbon sequestration potential of rangelands and grasslands. Records of ecosystem change are improving, mainly through remote sensing, but reliable data on land-use change are still fragmented and often not comparable – the extent of drylands, for example, is uncertain because of the classifications and methodologies used by different programmes. Similarly, there are discrepancies between a number of wetland inventories and there is no comprehensive global wetlands database.

Satellite remote sensing is an essential tool for monitoring global land resources, but no such technology exists for population patterns. National census efforts, the best current technique, are sporadic and underfunded in many countries, and there is a significant data gap for population changes in rural areas. Further, it is critical to track the consequences for the environment of rapid and extensive urbanization, with its uncertain implications for land resources.

Data on biofuels – including the extent of production and use – are incomplete at the global level, although national datasets can be found for some countries. Similarly, there is a need for improved national and global monitoring of land transactions including large-scale land deals. There are also few standard indicators that governments can use to monitor the environmental impacts of different patterns of land tenure. Finally, standard methodologies for the badly needed valuation of ecosystem services are at an early stage of development.

Goal Gaps

Issues of capacity building and stakeholder participation are also inadequately represented in international goals. Several of the land-related goals that do exist lack quantifable targets, complicating the task of assessing progress towards their achievement. A particular challenge is to acknowledge the interactions between different components of social-ecological systems at different scales.

Goals cannot be considered in isolation. Due to tensions and synergies, progress towards one goal must be viewed in light of implications for others. Meanwhile, efforts to address the education and health issues expressed in MDGs 2–6 can indirectly help achieve MDGs 1 and 7 in the long term. Thus, an integrated perspective on goal achievement is crucial.

Discussion of Key Issues Economic Growth and Land Resources

The global economy has quadrupled during the last 25 years, but 60 per cent of the world's major ecosystem goods and services underpinning livelihoods have been degraded or used unsustainably. This means that traditional economic growth cannot be the foundation of sustainable development. A new paradigm of economic welfare is required - one that is focused on improving human welfare and social equity, and reducing environmental risks and ecological scarcities. One such approach, the green economy proposed by UNEP in 2010, includes:

- valuation of natural resources and environmental assets;
- pricing policies and regulatory mechanisms that translate these values into market and non-market incentives; and
- measures of economic welfare that are responsive to the use, degradation and loss of ecosystem goods and services.

The transition from traditional economic growth to the green economy will require changes to national regulations, policies, subsidies, incentives and accounting systems, as well as to global legal and market infrastructures, an appropriate international trade structure and targeted development aid.

Meeting the Growing Demand for Food

Both global population and per-person consumption continue to grow. The achievement of MDG 1, the eradication of extreme hunger and poverty, will require getting more food to more people. How this is accomplished will have important implications for MDG 7 - environmental sustainability. Population growth is an important part of this complex interaction, but changing lifestyles and consumption patterns, particularly the increasing global demand for animal products, are also significant. Friction between these two MDG goals could be reduced by:

- improving efciency along the whole food chain by increasing crop yields through research and extension, and reducing food waste and spoilage by improving transport, storage and distribution infrastructure in developing countries and changing behaviour in wealthier societies, where much food waste occurs in food retail markets and homes;

- implementing full-cost accounting for food products that reffes the environmental and social costs of their production in order to facilitate a shift in consumption patterns;
- encouraging, where appropriate, innovative approaches to food production to shorten food supply chains and enhance food security;
- evaluating the ecosystem service and carbon balance implications of potential biofuel production to inform land-use planning and management, and reducing competition between food and biofuel production, particularly in areas with the highest crop production potential.

Growing Demand for non-food Resources

Crop- and plantation-based biofuel production has increased rapidly in recent years and the land-use transitions associated with this could have strong environmental and social impacts. Fuel-blending targets in numerous countries mandate the continued expansion of biofuel production. Next-generation biofuels - from, for example, algae or cellulose - are still under development and are not likely to contribute a significant share of biofuel production in the near future. Governments should recognize that targets for biofuel production have both direct and indirect implications for land use at national and global scales.

Large-scale land acquisitions are growing with potentially major impacts on land-use change and social relations. Recent reports have advocated the establishment of an observatory of land tenure and rights to food to monitor access to land and ensure that land investments result in decreased hunger and poverty in host communities and countries. United Nations organizations could play an important role in creating precedents that could help improve food access in developing countries.

Complexity and Policy Challenges

An important step towards addressing these challenges is to monitor, study and understand how social and biophysical drivers interact, and the diversity of social, economic and environmental consequences they generate at local, regional and global levels. A concerted effort by international organizations, the scientifc community, and national and local institutions could create the comprehensive monitoring network needed to achieve this goal – but to be effective there needs to be strong coordination between these actors.

Limitations in the assessment of land-change processes cannot and should not delay action to address their driving forces, with the precautionary principle being applied to reduce their negative impacts. Current evidence of their consequences highlights the need to act in the short term to avoid potentially irreversible negative outcomes in the long term. There are no easy answers to these complex problems, and single and isolated actions might achieve only limited positive outcomes rather than broad solutions. New governance approaches to land management could help incorporate adaptive management, capacity building and more efcient valuation of ecosystem services and natural resources by combining market-based tools with a bigger role for community agency and bottom-up approaches. New governance approaches could also help foster the changes in consumption patterns needed to reduce pressure on land systems and create better knowledge and awareness of the multiple values of ecosystems. While the leadership of UN organizations and other international institutions is a central element in these efforts, governments have a crucial role, responsibility and opportunity to act as agents of change.

References

ACIA (2005). *Arctic Climate Impact Assessment.* Cambridge University Press, Cambridge.

Bakker, M.M., Govers, G., Kosmas, C., Vanacker, V., van Oost, K. and Rounsevell, M. (2005). Soil erosion as a driver of land-use change. *Agriculture, Ecosystems and Environment* 105(3), 467481.

Corbera, E., Brown, K. and Adger, W.N. (2007). The equity and legitimacy of markets for ecosystem services. *Development and Change* 38(4), 587613.

FAO (2011). *2011: State of the World's Forests.* Food and Agriculture Organization of the United Nations, Rome.

Kumar, P. (ed.) 2010. *The Economics of Ecosystems and Biodiversity: Ecological and Economic Foundations.* Earthscan, Washington.

Montgomery, D.R. (2007). Soil erosion and agricultural sustainability. *Proceedings of the National Academy of Sciences of the United States of America* 104(33), 1326813272.

4

Global Water Governance

Aquatic ecosystems are major integrators of natural and anthropogenic processes. As the ultimate sink for pollutants, freshwater and marine ecosystems are among the most sensitive indicators of the environmental impacts of human activities. They support a wide diversity of life, providing important goods and services that directly or indirectly support and sustain human existence and livelihoods. Adequate freshwater supplies of acceptable quality are recognized as a human right by the UN General Assembly's declaration on clean water and sanitation.

As highlighted in the *Millennium Ecosystem Assessment,* freshwater and marine ecosystems provide various services, including provisioning (food, water, fbre, fuel), regulating (climate, hydrological, purification), cultural (spiritual, recreational), and supporting (sediment transport, nutrient cycling). Such ecosystem services are a function of water, land, biodiversity and atmospheric links. Healthy aquatic ecosystems not only provide goods and services, but also enhance resilience against the negative impacts of environmental perturbations or disasters. Aquatic systems also drive major global bio-geochemical cycles; the open oceans play a major role in regulating global climate and weather patterns.

It will addresses freshwater and marine systems as distinct but linked hydrological components of the water environment. It assesses progress towards achieving water-related goals in major multilateral environmental agreements identified by the *GEO-5* High-Level Intergovernmental Advisory Panel and regional consultations. Based on the drivers-pressures-states-impacts-responses framework (DPSIR) used for the *GEO-5* assessment, this focuses on the state, trends and impacts of the water environment, with references to drivers and responses, and other environmental sectors where appropriate.

Although freshwater ecosystem goods and services are extensive, competition and multi-sectoral demands for water have resulted in overexploitation and contamination of resources in many regions. Competing water uses and their impacts on sustainable aquatic resources, including quantity and quality issues, are discussed, including the water needs of ecosystems. It also addresses inequitable and unsustainable water demands in many countries. Pollution from land- and marine-based activities continues to degrade coastal areas and open oceans. Water quality trends are discussed. Continuing overfishing severely impacts many fish stocks, particularly marine species.

Many predicted global climate change impacts will be manifested in changes to the hydrologic cycle. How they could affect the water environment is highlighted, including increased frequency, duration and severity of droughts and foods. Predicted climate change impacts and their uncertainties are discussed, including the vulnerabilities and adaptation needs of many communities.

Watersheds comprise a group of linked water systems that can include rivers, lakes, reservoirs, wetlands, underlying aquifers and downstream marine systems, although these links are often not considered in developing water management plans. This is a significant omission, since simultaneously managing for human health and social concerns – including disease and poverty, economic development and sustainable environmental integrity – within complex global connectivity often requires environmental and economic trade-offs, some very difficult. Because many water-related problems result from policy, institutional, financial or other governance inadequacies, this will also discusses both freshwater and marine governance elements identified in the multilateral environmental agreements.

Internationally Agreed Goals

Freshwater was selected as a priority issue in all UNEP *GEO-5* regional scoping consultations, with most regions identifying Paragraph 26c of the Johannesburg Plan of Implementation as the most important freshwater goal, with water availability and marine issues also identified in several regions.

Goals were identified on the basis of their policy relevance and ability to illustrate intergovernmental cooperation since the United Nations Conference on Environment and Development (UNCED) in 1992 and earlier.

Improve the efficient use of water resources and promote their allocation among competing uses in a way that gives priority to the satisfaction of basic human needs and balances the requirement of preserving or restoring ecosystems and their functions, in particular in fragile environments, with human domestic, industrial and agriculture needs, including safeguarding drinking water quality.

Water Scarcity

Water scarcity is a significant and increasing threat to the environment, human health, development, energy security and the global food supply. Ecosystems, which provide life-supporting goods and services, sufer from multiple pressures, including the need for water of adequate quantity and quality as well as appropriate timing (environmental fows). The indicator used here is blue water scarcity, the proportion of groundwater and surface water consumed relative to the sustainable water available for human use, after accounting for environmental fows. Water scarcity is a significant factor in human water security, with a ffifth of the global population living in areas with physical water scarcity.

Falkenmark and Rockström estimated the water required to maintain ecosystem goods and services as 75 per cent of the total water use, while direct human water use represented 25 per cent of the total. These figures include both blue (groundwater and surface water) and green water (water stored in the soil). Water is overcommitted in many places, leaving insufcient resources for both human and environmental needs. In a study of 424 of the world's major river basins, containing a population of 3.9 billion people, environmental fow requirements were violated in 223 basins, containing 2.67 billion people facing severe water scarcity during at least one month of the year. Although arid regions of Northern Africa and the Middle East are not included in this analysis, other data suggest that the proportion of renewable water withdrawn in those regions exceeds 50–75 per cent, leaving little environmental fow.

Although many goals in the Johannesburg Plan of Implementation acknowledge the importance of marine and coastal ecosystems, there is less recognition of water needs to support freshwater ecosystems, which are themselves legitimate water users. Although the importance of formally recognizing the environment as a legitimate water user is increasing, it remains on a relatively small scale in practice, with many aquatic ecosystems still at risk.

Water Demand

Global withdrawals have tripled over the last 50 years to meet the demands of a growing population with increasing wealth and consumption levels. While water supply over this period has remained relatively constant, demand now exceeds sustainable supply in many places, with serious long-term implications (2030 Water Resources Group 2009). The planetary boundary for human consumptive blue water use -when used groundwater and surface water is not made available for reuse in the same basin - is estimated to be 4 000 km^3 per year, with current consumptive blue water use estimated at approximately 2 600 km^3 per year. Projected water demands may to reach planetary boundaries in the coming decades.

Agricultural, industrial and domestic water withdrawals have steadily increased. Agriculture is by far the largest global water user, with withdrawals for this purpose being unsustainable in many places due to unbalanced long-term irrigation water budgets, as evidenced by the mining of aquifers and reliance on large water diversion projects. These withdrawals are projected to continue increasing, placing further pressure on aquatic ecosystems, which themselves also require water of adequate quantity, quality and timing for sustained health.

Many communities are dependent on unsustainable groundwater withdrawals (aquifer mining) to meet agricultural and domestic water demands, further threatening water security in many regions. Between 1960 and 2000, global groundwater withdrawal increased from 312 km^3 to 734 km^3 per year, resulting in groundwater depletion increasing from 126 km^3 to 283 km^3 per year. Many globally important agricultural centres are particularly dependent on groundwater, including northwest India, northeastern China, northeast Pakistan, California's Central Valley, and the western United States.

Not all water withdrawals result in consumptive water use, since much withdrawn water is returned in the form of wastewater or irrigation return fows. Rain-fied agriculture also represents significant human water use without direct water withdrawals. Global water consumption per person, as measured by the water footprint, averages 1 387 m^3 per year. North America has the highest water footprint at 2 798 m^3 per person per year, while Asia and the Pacifc have the lowest at 1 156 m^3 per person per year. Of the total global water footprint, 74 per cent represents rainwater stored in soil (green water), 11 per cent represents the consumptive use of surface and groundwater (blue water), and 15 per cent represents the freshwater required to assimilate pollution from all sources (referred to in the water footprint terminology as grey water). Agriculture accounts for 92 per cent of the total global water footprint; livestock and related products alone account for 27 per cent.

Water use Effciency and the Virtual Water Trade

Since the renewable supply of water is relatively constant, addressing water scarcity relies in large part on reducing water demand by improving efciency and reducing consumptive water use. All user demands must be considered together, including environmental water requirements.

Although improved methods and technologies have produced efciency gains in all sectors in some regions, the need, and potential, exists for further improvements to ensure the well-being of a growing world population while minimizing the impacts on ecosystems and their goods and services.

The need and potential for improvement is greatest in the agricultural sector, since approximately 70 per cent more food will be needed by 2050 to cope with a growing population and dietary changes. Improvements in the application, conveyance, distribution and management of irrigation can raise the overall efciency of the water getting to the crops from approximately 35 per cent to 75 per cent or more. Broader agricultural water-use efciency strategies include land-water management and reuse, while food supply chains beyond the farm can also be more water efcient.

There is inadequate global data to evaluate the overall state and trends of industrial and domestic water-use efciency. Nonetheless, opportunities do exist for significant improvement in these sectors, particularly where there are major withdrawals and/or rapid urbanization is occurring. Water allocation efciency is also required at the river-basin level to ensure sustainable, equitable and economic water use.

At national, regional and global scales, the virtual water trade – the water embedded in traded products ranging from crops to manufactured goods – can be a tool for improving overall efciency by capitalizing on the comparative advantages of certain water uses in particular regions. About one-ffifth of the global water footprint is related to production for export. The global virtual water trade for agricultural and industrial products totalled 2 320 km^3 per year between 1996 and 2005, with crops contributing 76 per cent and animal and industrial products each contributing 12 per cent. The virtual water trade can efciently redistribute water and partially help to address the disconnection between consumption and the impacts of production. Water-scarce basins, countries or regions, for example, can import water-intensive products through trade, preserving scarce water resources for more valuable purposes. However, it can also lead to overexploitation of water resources in net exporter countries, prioritizing commodity water needs over basic local needs, especially where strong economic drivers promote commodity exports. Another characteristic is that some net virtual water exporters, such as Australia or South Asia, are also water scarce, while some net importers may have abundant supplies, as is the case for Central Europe.

Changes to the Hydrologic Regime Extreme Events

The number of food and drought events classified as disasters – when ten or more people are killed, 100 are affeed, a state of emergency is declared or international assistance is requested – has risen since the 1980s, as have the total area and number of people affeed and the level of damages. River channelization, foodplain loss, urbanization, particularly in coastal areas, and changing land use are major reasons for the increasing impacts of foods and droughts as well as growing vulnerability to those impacts. The number of people affeed and total damages vary significantly, making it hard to identify trends with confdence. Vulnerability depends on the preparedness and capacity to anticipate and react to extreme events. There are varying preparedness levels on a regional basis for dealing with sudden-onset (foods) and gradual-onset (droughts) disasters.

Floods cause loss of life and billions of dollars of damage annually, with the economic losses higher in developed countries due to the financial valuation and insuring of assets. Between the 1980s and the 2000s, a 230 per cent rise in the number of food disasters was accompanied by increasing levels of damages. In addition, the number of people exposed to foods increased by 114 per cent. Over 95 per cent of deaths related to natural disasters between 1970 and 2008 occurred in developing countries and although governments in South and East Asia, for example, increased their disaster preparedness levels, the capacity of communities to cope with such extreme events is weakening because of inadequate social capacity and greater food severity. Looking to the future, higher precipitation intensity is forecast for the northern hemisphere and equatorial areas, with many already arid and semi-arid areas expected to get drier.

The number of drought disasters rose by 38 per cent between the 1980s and the 2000s, the number of people affeed increased and related damages also increased. Droughts disrupt sustainable social and economic development, hindering achievement of the Millennium Development Goals (MDGs), and place additional stresses on ecosystems. Communities dependent on rain-fied crops, which represent approximately 70 per cent of global crop production, often have few alternative food sources beyond international aid. This is evidenced by the severe ongoing drought in Eastern Africa and the reduction of net primary production in Latin America, Africa and South East Asia. Droughts also affect irrigation and can exacerbate water resource conficts, with arid and semi-arid areas being particularly vulnerable, especially in the context of climate change.

Dams and River Fragmentation

Dam building and river control significantly benefit humans, providing food protection, reliable water supplies and hydroelectric power. But dams can also have detrimental impacts on ecosystems, including channel fragmentation and fow modification, altering ecosystem processes and affeing aquatic organisms, particularly migratory species. Improved management of existing dams to ensure environmental fows and retain or create fish passes is important to mitigate conficts, although such measures often fall short as a full remedy. Careful trade-of analyses are necessary to ensure that the design, location and operation of new dams minimize environmental impacts.

Dam density is highest in industrialized countries, although construction in developed regions has slowed because most suitable locations have already been used, and because recent legislation and public pressure do not support dam construction. However, dam building is being actively pursued in many developing countries to secure water and electricity supplies. As this trend is likely to continue, dam planning should consider any predicted increase in fow variability associated with climate change.

Groundwater Contamination

Groundwater around the world is threatened by pollution from agricultural and urban areas, solid waste, on-site wastewater treatment, oil and gas extraction and refning, mining,

Regions of Greatest Concern

Arsenic is of particular concern in Bangladesh, India, highly populated river deltas in South East Asia, North America and Eastern Europe manufacturing and other industrial sources. The primary causes are inadequate control of these activities and exceedance of the natural attenuation capacity of underlying soils and strata. Salinization of overexploited aquifers, especially in coastal areas, is another serious concern, particularly for communities dependent on groundwater for drinking.

Groundwater nitrate concentrations are increasing, especially in areas of rapid urbanization, inadequate sanitation and/or heavy agricultural fertilizer use. Nitrate in groundwater contributes to eutrophication and has direct human health impacts. Both naturally occurring arsenic and that mobilized by human activities threaten drinking water quality in many countries. Groundwater contaminated with arsenic from natural geologic sources affes 35–75 million people. Surface water pollution in some regions has led to the development of groundwater as a source of drinking water, resulting in inadvertently exposing people to these natural sources of arsenic.

Pathogenic Contamination

Pathogenic contamination of surface and groundwater is a critical threat to human health in many areas and contributes to water treatment costs in many communities. Using domestic sewage collection and treatment as a proxy, microbial contamination has decreased over past decades in most developed countries. In contrast, microbial pathogens are often the most pressing water quality issue in many developing countries.

Because human and animal faeces are the primary pathogenic sources of water contamination, achieving MDG Goal 7c of halving the population without basic sanitation by 2015 will help reduce such pollution. Nevertheless, although some regions have made significant progress, the world is currently not on track to attain this goal. Improved sanitation continues to bypass the poorest communities and individuals, especially in Africa and South Asia. Unless achieving the MDG sanitation goal in the future includes the provision of wastewater collection and treatment facilities, increasing access to improved sanitation could have the unintended negative impact of delivering more untreated wastewater to water bodies, further degrading downstream water quality.

Nutrient Pollution and Eutrophication

Eutrophication, resulting from excessive nutrient pollution from human sewage, livestock wastes, fertilizers, atmospheric deposition and erosion, is a continuing, pervasive water quality problem. Although there has been increased sewage treatment in many areas, much less progress has been made in reducing nutrient loads from non-point sources, including agricultural and urban run-of and atmospheric deposition to freshwater and marine systems. Interference with global nutrient cycles may be reaching planetary boundaries, beyond which marine and freshwater ecosystems might not recover, although specifc thresholds for these processes remain uncertain.

Global river nutrient export has increased nutrient export has increased by approximately 15 per cent since 1970, by approximately 15 per cent, since 1970, with South Asia accounting for at least half of the increase. There has been a 74 per cent increase in algal and macrophyte gross productivity in lakes since 1970, and a dramatic increase in the number of eutrophic coastal areas since 1990. Under severe eutrophic conditions, algal blooms can produce hypoxic conditions, causing fish kills in lakes, and dead zones in coastal areas.

Hypoxia has become a significant and increasing problem in lakes and rivers, estuaries and coastal areas around the world. At least 169 coastal areas are considered hypoxic, with dead zones especially prevalent in the seas around South East Asia, Europe and eastern North America. Only 13 coastal areas appear to be recovering, most in North America and northern Europe. Whereas phosphorus loads are projected to level of, global river nitrogen loads are likely to increase by an additional 5 per cent by 2030, mostly in South Asia.

Nutrients can also cause harmful algal blooms in freshwaters and coastal areas, some releasing algal toxins that directly affect human health, aquatic organisms and livestock. The number of reported outbreaks of paralytic shellfish poison, a harmful algal toxin found in eutrophic waters, increased from fewer than 20 in 1970, to more than 100 in 2009.

Marine Litter

Litter is found in all the world's oceans because of poor solid waste management and the increased use of plastic. It damages wildlife, fisheries and boats, contaminates coastal areas, and presents safety and human health risks. Marine litter accumulates on coastal beaches, on the sea bottom and large marine gyres in both the Atlantic and Pacifc Oceans.

Of the 12 seas surveyed between 2005 and 2007, the South East Pacifc, North Pacifc, East Asian Sea and Wider Caribbean coasts contained the most marine litter, and the Caspian, Mediterranean and Red Seas the least. Regional studies of the Baltic Sea, Northeast Atlantic, US coastline and North Atlantic Subtropical Gyre indicated no statistically significant changes in litter quantity between 1986 and 2008, while data from the Mid-Atlantic indicated an increase in land-based and general-source marine litter during 1997–2007.

Persistent Toxic Chemicals

Toxic pollutants include the trace metals cadmium, lead and mercury, pesticides and their by-products such as dichlorodiphenyltrichloroethane (DDT) and chlordecone, industrial chemicals and combustion by-products. They are still used in many places and thus continue to accumulate in aquatic systems, leaving a legacy of sediment contamination; they are found in 90 per cent of water bodies. The pollutants of greatest concern are persistent, toxic and bioaccumulative. Organisms can accumulate contaminants from water, sediment and food, acquiring tissue contaminant levels much higher than those in the surrounding environment. Organochlorine compounds such as polychlorinated biphenyl (PCB) or DDT concentrate in fatty tissues, remain for long periods and biomagnify up the food chain, with the highest concentrations found in top predators.

Concentrations of many persistent organic pollutants (POPs), which tend to accumulate in the Arctic, have decreased in Arctic air samples since the early 1990s. Tissue concentrations of at least three organochlorine chemicals in 12 deep-sea fish species in the Western Pacifc and PCB concentrations in at least four fish species in San Francisco Bay have also fallen since the mid-1990s.

Emerging Water Quality Concerns

Although conventional toxic pollutants are declining in many industrialized areas, additional contaminants are raising new concerns, for example the use of fame retardants such as polybrominated diphenyl ethers (PBDEs), a type of POP, has increased exponentially over the past 30 years in Europe, North America and Japan. There are also mounting concerns about pharmaceuticals and personal care products that are not removed by most sewage systems, and thus enter the environment after use. The long-term risks to aquatic organisms and humans are largely unknown, although it is clear that pharmaceuticals and endocrine-disrupting compounds can have biological effes at very low concentrations.

Nanoparticles and microplastics are relatively new water pollutants. Nanoparticles – particles measuring 1–100 nanometres, or billionths of a metre – are increasingly used in modern life. An emerging feld of nanoecotoxicology is examining their environmental fate and potential impacts on aquatic ecosystems. Microplastics, from the deterioration of plastic objects, may contain additives that accumulate in aquatic organisms, and their concentrations, especially in marine systems, are expected to follow increases in global plastic consumption. Additional types of pollutants about which little is currently known will doubtless continue to be identified. Although not new, industrial, medical, military and accidental releases of radioactive substances are of renewed concern, as illustrated by the water contamination after the 2011 tsunami damaged Japanese nuclear power plants. Invasive alien species also remain a problem for many coastal areas.

Cross-cutting Issues

Water Security and Human Health

As previously noted, regional diferences exist regarding both absolute water availability and the limitations placed on it by inadequate infrastructure. Both relate to water security, as does water pollution, because they can all affect the range of human and environmental water uses. Despite improvements, lack of access to drinking water of adequate quality and quantity remains one of the largest human health problems globally. Inadequate water supply is an inherently regional phenomenon, however, caused by basin-level water scarcity, regional water quality, inadequacies of infrastructure and governance, cultural perspectives and inequitable water pricing.

Water Security

Although several defnitions for water security have been proposed since the

1992 Rio Earth Summit, none has been universally accepted. Varying defnitions, leading to numerous indices based on different criteria, make it difficult to generate trend data. The Ministerial Declaration of the Hague broadly defnes water security to include the protection and improvement of freshwater and marine ecosystems, sustainable development and political stability, with the aim of providing every person with access to "enough safe water at an afordable cost to lead a healthy and productive life", as well as protecting vulnerable communities from water-related risks and hazards.

About 80 per cent of the world's population lives in areas with high water security threats, the most severe category encompassing 3.4 billion people, almost all in developing countries. Water security threat here refers to the cumulative effect of 23 drivers that have an impact on water resources, categorized into watershed disturbance, pollution, water resource development and biotic factors. More people are likely to experience severe water stresses in the coming decades because of increased demands in addition to altered precipitation patterns associated with climate change.

With higher investments in infrastructure in the industrialized countries, the figures show that human water security can be increased, overcoming the various threats to water resources, while low investments in developing countries means their water security remains poor. Investments must be coupled with adequate institutional capacity, and because infrastructure development often occurs at the expense of aquatic biodiversity and environmental quality, it is imperative that environmental risks related to investments are considered and appropriately mitigated.

Equitable Access to Improved Drinking Water

Although water security is an increasing problem in many regions of the world, significant progress in access to improved drinking water has been made since 1990. However, several regions, including most of Africa and other rural areas in developing countries, still lack access to improved drinking water sources. The UN General Assembly declared access to clean water and sanitation as a human right in July 2010, although the right is not yet recognized or applied in many countries.

Recent data suggest that the MDG drinking water target was met in 2010. However, there are important inequities in this improvement. Whereas only 4 per cent of people in urban areas lacked access to improved drinking water in 2010, in rural

At any given time, over half the world's hospital beds are filled with people suffering from water-related diseases. Diarrhoeal diseases make up more than 4 per cent of the global disease burden, 90 per cent being linked to environmental pollution and lack of access to safe drinking water and sanitation. Africa has the highest burden of diarrhoea-related childhood deaths, accounting for 70 per cent of the 1.3 million deaths of children less than five years old in 2008. Not surprisingly, access to basic sanitation is also poorest in sub-Saharan Africa, with 330 million people lacking access to proper sanitation.

Such diseases are a major public health concern, especially in Africa. Globally, diarrhoea related to inadequate sanitation and water supply was the second largest contributor to the 2004 global disease burden, claiming more than 70 million disability-adjusted life years (DALYs), years lost due to ill-health, disability or early death. Global health statistics indicate that Africa and South Asia contain the areas most severely affeed by waterborne disease.

The WHO is focusing on reducing 25 different water-related diseases. There have been some notable successes in the reduction of onchocerciasis, malaria, schistomiasis and cholera. However, globally reported cholera incidence – which serves as a proxy where complete data on water-related disease trends are lacking – has increased in recent years, mainly in Africa. In 2009, 45 countries from all continents reported 221 226 cases of cholera. Water-related diseases are a continuing public health problem in developing countries lacking access to adequate drinking water and sanitation, as further evidenced by the cholera epidemic in Haiti following the 2010 earthquake.

Water-energy-climate Nexus

Water, energy, economic development and climatic change are interdependent issues. Increases in human population and per-person consumption related to economic development drive energy demands. Meanwhile the use of fossil fuel energy produces greenhouse gas emissions that contribute to climate change, which has effes on water, including extreme weather events, loss of ice cover, water scarcity and sea level rise. In turn, responses to climate change have implications for the water environment. Some forms of solar energy consume significant quantities of water, US$9–11 billion per year, 85 per cent of this in often in arid regions. With increasing water scarcity, some regions developing countries. There are additional costs associated with food also rely on desalinization of marine water, requiring large energy inputs. In addition, droughts have the potential to decrease hydropower production.

Climate Change Impacts on The Water Cycle and Ocean Warming

The hydrologic cycle refers to the continuous movement of water through the oceans, atmosphere and over and under land surfaces. There is strong evidence that climate change is altering global and regional hydrologic cycles, with impacts predicted to be manifested as changing precipitation patterns, increased intensity of extreme weather events and consequent natural disasters, retreating glaciers resulting in altered river discharge regimes, and more intense droughts in semi-arid regions.

Although there is considerable uncertainty regarding projected impacts on specifc water systems, climate change has the potential to seriously affect water management. Nonetheless, the global impacts of other human activities on the hydrologic cycle – urbanization, industrialization, water resources development – are likely to exceed those related to climate change, at least for the next two to three decades. If climate change impacts are to be addressed, the cost of the additional water infrastructure needed by 2030 to provide a sufcient quantity of water for all countries is estimated at risk management and water quality protection. There are signs of increased awareness of mitigation and adaptation needs: of 191 water projects funded by the World Bank between 2006 and 2008, 35 per cent incorporated mitigation and adaptation measures for climate change impacts. At the same time, however, local and regional efforts to increase protection against foods and other extreme events are likely to have significant negative impacts on aquatic ecosystems themselves.

The most direct climate change impact on oceans is increased sea surface temperature (SST), which has risen by 0.5p C globally since the 1980s and is predicted to continue increasing throughout the 21st century. Global precipitation is predicted to increase at a rate of 1-3 per cent per degree of surface warming, with more extreme precipitation events predicted for many tropical and temperate regions.

Melting Ice Sheets and Sea Level Rise

Sea level rise is caused by ocean thermal expansion and by melting glaciers and ice sheets. Although average global sea level has remained relatively constant for almost 3 000 years, it increased by approximately 170 mm during the 20th century, and is projected to rise by at least another 400 mm (+/-200 mm) by 2100. Measurements from 1993 to 2008 indicate that sea levels are already rising twice as fast as in previous decades and are exceeding the rise predicted by climate models.

Although there is considerably variability associated with these and other estimates of sea level rise, 25-50 per cent of sea level rise observed since 1960 has been attributed to thermal expansion. Some variability may result from water impounded in reservoirs, which is estimated to have reduced sea level rise by 30–55 mm over the past 50 years. Small glaciers and ice caps exhibited significant mass losses over the 20th century and freshwater run-of from melting land-based ice sources will increase in the future. However, with losses accelerating over the past 20 years, melting Greenland and Antarctic ice sheets have become the biggest contributors to sea level rise, and will remain the dominant contributor to sea level rise in the 21st century if current trends continuc.

Because of the high concentrations of human populations and infrastructure in coastal zones, many countries are vulnerable to sea level rise and associated coastal and low-lying community fooding. Developing countries, particularly small island developing states (SIDS) and deltaic areas, are especially vulnerable, many with limited capacity to adapt to rising sea levels or recover from associated losses. The estimated costs of coastal adaptation range from US$26 billion to US$89 billion per year by the 2040s, depending on the magnitude of sea-level rise.

Ocean Acidification

The oceans annually absorb a substantial proportion of anthropogenic carbon dioxide (CO_2), which reacts with water to form carbonic acid, thereby making the ocean more acidic. The mean surface ocean pH has already decreased from a pre-industrial average of about 8.2 to a present value of 8.1, though there are regional diferences, and Feely *et al.* project a pH decrease to a mean of about 7.8 by 2100. Ocean acidification may be approaching the planetary boundary.

Increased ocean acidity affes marine animals with carbonate shells and skeletons, calcareous algae and other organisms. Afected organisms include reef-building corals as well as animals critical to ocean food webs, including several important human food sources such as crabs and molluscs. Combined with higher water temperatures, ocean acidification is thought to be a major cause of coral bleaching, destroying coral reef ecosystems around the world, with some studies projecting a rapid contraction of tropical coral reefs by 2050. Coral reefs provide important ecosystem services, such as spawning and nursery grounds for some commercially important fish species. Impairment of these ecosystems and their services are becoming increasingly evident and illustrate the need for governance to enhance their protection.

Impacts of Energy Development on Water Resources

While global data are lacking, the energy sector is believed to account for approximately 40 per cent of total water withdrawals in the United States and European Union (EU). Water demands for energy range from extraction and processing of raw materials to driving hydropower turbines and cooling thermoelectric plants, including nuclear. Fossil fuel extraction can also have serious impacts on water quality.

Oil and gas exploration and production can affect both freshwater and marine ecosystems. Newly proven technologies are accelerating the expansion of new natural gas wells in shale gas basins. Associated water resource impacts are currently being researched, including aquifer contamination with potentially explosive methane levels, surface and groundwater contamination, streams receiving water discharges, and high consumptive water use for well drilling and completion. Oil sand exploitation also requires large water volumes and can produce severe water pollution.

Oil spills continue to pose an environmental threat, particularly to marine ecosystems. Although the number of oil tanker spills has decreased significantly since the 1970s and 1980s, the recent large spill associated with offishore oil and gas exploration in the Gulf of Mexico is evidence of ongoing risks to marine ecosystems. Nevertheless, with increasing global oil and gas demands, such offishore activity is expected to increase over the next two decades, facilitated by the resolution of maritime boundaries and improved access to previously inaccessible areas as Arctic ice melts. The Arctic contains approximately 20 per cent of the world's undiscovered but technically recoverable oil and gas resources, but the region is uniquely vulnerable to oil spills because of its remoteness, harsh physical environment, the aggregation of large numbers of marine mammals and the slow rate of oil degradation in cold water.

The most water-intensive form of electricity production is biomass, followed by hydropower, oil, coal and nuclear, gas, some concentrated solar power systems and geothermal, solar photovoltaics, and wind. Exact values vary greatly, depending on electricity generation type and location. Many forms of concentrated solar power, for example, which may be most effective in arid areas exhibiting high solar energy levels, also require significant quantities of water for cooling, sometimes as much as fossil-fuel-powered plants. There are cases in which water scarcity is already affeing energy production. More than half of existing or planned capacity for major power companies in South and South East Asia, for example, is located in water-scarce or water-stressed areas.

Climate change mitigation policies can also affect water demands for electricity production. Capture and storage of carbon emissions from coal-fred plants, for example, can increase water consumption by 45–90 per cent. Further, increasing the proportion of electricity generated from biomass or some types of concentrated solar power is likely to have significant negative impacts on water availability, highlighting the need for selecting power generation types that use less water and more efcient technologies.

Water Governance

Agenda 21 of UNCED called for "integrated approaches to the development, management and use of water resources", subsequently leading to the development of several integrated management paradigms, including integrated water resources management, integrated lake basin management, and integrated coastal zone management, as mentioned in the Jakarta Mandate on Marine and Coastal Biodiversity and other outputs of the Convention on Biological Diversity (CBD). Integrated management approaches also ofer a degree of protection against the negative impacts of natural disasters such as the devastating earthquake and tsunami that struck Japan in 2011.

The need for integrated approaches was formalized in Paragraph 26 of the Johannesburg Plan of Implementation, which states that governments should develop integrated water resources management and water efciency plans by 2005 through actions at all levels. This overall target has not been met. However, data from studies in 2003 and 2005, primarily involving developing countries, and for 2008 and 2012 for all countries, do suggest significant headway – from the development of plans to their implementation

– particularly in developed countries. Progress appears to have slowed, however, in developing countries.

Some water professionals and policy makers suggest that in some cases the integrated management concept is insufciently specifc for practical implementation, and is slow to get under way owing to many institutional, economic, political and resource constraints. Further, although integrated water management proposes cross-sectoral coordination, all relevant government agencies and key water stakeholders may not be in agreement. Because participatory approaches do not always ensure consideration of gender perspectives, a systematic assessment of the difering impacts of economic development on women and men is also necessary to ensure that water issues affeing both genders are part of programme planning, implementation and evaluation, including ensuring institutional and organizational changes for gender equality as an ongoing commitment. Further, although integrated management can be applied at many levels, from village to basin to national to transboundary, there are management issues particular to each of these levels, necessitating both bottom-up and top-down approaches. Existing evidence, however, suggests that integrated policies have focused largely on higher-level activities such as national policy reforms or the establishment of river basin organizations, rather than on on-the-ground implementation of integrated management activities at the local level.

The European Commission (EC) applied integrated water resources management principles in its Water Framework Directive in 2000, and a food risk management directive in 2007. Further, although implicit in goals such as Paragraph 26 of the Johannesburg Plan of Implementation, there is no global multilateral environmental agreement specifically directed at aquifer conservation. There are, however, several regional groundwater initiatives, including the 2008 establishment of the Africa Groundwater Commission. Because poor groundwater governance is a major issue, recognizing groundwater systems in national laws would be a first step towards improved groundwater governance, followed by establishing sustainable institutions and fnancing.

While social science literature on regional experiences with integrated water management and international river basin management is increasing, there is little data on the state and trends of such approaches, particularly their long-term benefits and impacts. Research has focused more on the concept and its application, and less on relevant policy implementation, highlighting the need for better progress indicators as well as continued monitoring frameworks to assess effectiveness. Certain policy initiatives have supported, *inter alia*, international river basin regimes, and include the Transboundary Water Assessment Programme (TWAP), to be implemented by UNEP for the purpose of developing a methodology for monitoring and assessing trends regarding, among other things, environmental and human water stresses, pollution, population density and water system resilience.

Marine Governance

Marine systems are a major food source, a means of transport for international shipping, a tourism attraction and a climate change regulator. Coastal dunes and tidal wetlands are important bufers against tidal fooding. A number of international conventions have been established to protect the marine environment, demonstrating significant levels of international cooperation, although a common limitation is their dependence on national legislation that may reffe other agendas.

Regarding international agreements, the 1972 Convention on the Prevention of Marine Pollution by Dumping of Wastes and Other Matter (the London Convention) and the 1973 International Convention for the Prevention of Pollution from Ships (MARPOL) address marine pollution. The United Nations Convention on the Law of the Sea, ratified by 160 countries and in force since 1994, represents a unified approach towards shared use of the oceans and their resources, addressing navigation, economic rights, pollution, marine conservation and scientifc exploration. Notwithstanding the concern with increasing marine litter, the conventions are generally viewed as positive frameworks for controlling and preventing marine pollution. The 2004 International Convention for the Control and Management of Ships' Ballast Water and Sediments exemplifes collaborative actions to address the introduction of alien invasive species, which can cause significant environmental and economic damage.

Another noteworthy international effort is the Global Programme of Action for the Protection of the Marine Environment from Land-based Activities (GPA), adopted by 108 governments and the EC in 1995. Although not enforceable, the GPA was designed to guide national and regional authorities in undertaking sustained action to prevent, reduce and/ or eliminate marine degradation from land-based activities. Many countries profess to subscribe to its goals, providing a means for developing collaborative strategies to address coastal and offishore water degradation from influent freshwaters. Marine spatial planning, similar to land planning or zoning on public lands, is another emerging area of possibilities for marine governance.

The regional seas conventions (UNEP and independent conventions), other action plans, and the large marine ecosystem concept promulgated by the US National Oceanic and Atmospheric Administration (NOAA), also represent integrated management approaches. Development and implementation of these plans difer, however, depending on the countries involved, with some programme guidelines being binding on the participating states while others are not.

The open oceans beyond national jurisdiction comprise almost half the planet's surface, with rapidly advancing technology opening major new oceanic frontiers for commercial uses including fishing, shipping, resource exploration and for potential marine engineering such as deep-ocean CO_2 sequestration. The open oceans and deep-sea ecosystems, including seamounts, trenches and canyons, cold water corals and hydrothermal vents, exhibit a wealth of biodiversity. Larger, slow-growing, long-lived and heterogeneously distributed species are adapted to stable conditions in these environments, and are particularly sensitive to environmental stresses. Governance of areas beyond national boundaries is weak and fragmented, however, and requires strengthening in preparation of increasing human activities and their impacts on areas within national jurisdictions, as well as to ensure conservation and sustainable use of the open oceans.

Water as a Basis for Confict and Cooperation

Competition for shared water resources can cause conficts, particularly at the local level, with water needs usually immediate, and resources often inadequate to address all competing needs. Intrastate conficts occur between rural and urban sectors – such as agricultural, industrial and municipal – and between water-reliant livelihood activities – such as fishing, agriculture and livestock grazing. Population growth, economic development and climate change can exacerbate management issues. Further, about 40 per cent of the global population lives in transboundary river basins that cover nearly half of the Earth's land surface and provide more than 60 per cent of global freshwater fows, imparting an additional management difficulty.

There also are increasing incidents of deliberate or threatened poisoning of water supplies. Sixty-nine water conficts were documented for 2000– 2010 in the Water Confict Chronology List maintained by the Pacifc Institute, compared to 54 recorded for 1975–1999. Although specifc incidents were not described in detail, De Stefano *et al.* found about 67 per cent of 1 831 reported water events for 1948–1999 were cooperative, with only 28 per cent being confictive in nature; by 2000–2008, the proportion of confictive events had increased slightly, to 33 per cent. Infrastructure and water quantity were consistently the major issues likely to generate confict. Although water conficts have occurred in many locations, and disputes could increase in the future, current evidence suggests greater potential for cooperation than for confict, particularly at the international level.

About 158 of the 263 international freshwater basins still lack cooperative management frameworks, while less than 20 per cent of the 106 basins with water institutions have multilateral agreements in effe. Evidence suggests, however, that freshwater systems with established transboundary basin organizations can usually improve cooperation, major examples including the Lake Victoria basin and the La Plata, Mekong and Senegal River basins. In fact, about 295 international water agreements have been signed since 1948. There are relatively few transboundary groundwater institutions, though codification of the law of transboundary aquifers by the UN International Law Commission (ILC), adopted by resolution of the UN General Assembly in 2008, is a major advance.

While transboundary basin organizations have facilitated so-called hydro-diplomacy, confict management and dispute resolution, there also are contrary examples. Although water scarcity in the Senegal River basin resulted in cooperation, the subsequent building of dams precipitated violent confict. Further, with population growth and climate change, water scarcity may lead to new confict constellations, including climate-induced degradation of freshwater resources, declining food production and increased storm and food disasters, which may further undermine food security.

The sustainable use of coastal areas and ocean resources requires effective coordination and cooperation at regional and global levels, with examples including UNEP's 13 regional seas programmes and the 64 large marine ecosystems. The EU Marine Strategy Framework Directive is another regional instrument, applicable in European waters under the jurisdiction of EU Member States that border the Baltic, Black and Mediterranean Seas and North-East Atlantic. Although the regional seas programmes and large marine ecosystems are consistent with UNCLOS and generally reffe the targets of the 2002 World Summit on Sustainable Development (WSSD), their achievement status remains unclear.

Outlook and Gaps

There has been progress since 1990 in achieving goals directly related to human well-being and economic development, including access to water supplies and reduction of some toxic pollutants threatening human health. Water-related diseases and water supply in the rural areas of developing countries, however, require increased attention. There has also been progress on water governance with the development of integrated water resource management plans and transboundary water agreements. However, these plans must now be implemented, adequately funded and enforced to improve aquatic ecosystems and the sustainability of their life-supporting goods and services.

Improving water security and ensuring equitable access to water resources remains a challenge. Against a background of continuing water degradation and overexploitation, the need for sustainable water supplies remains one of humanity's most critical resource needs. There also has been little to no progress in most regions in reducing nutrient loads to freshwaters and coastal areas, or for governance beyond national jurisdictions.

The complexity of the drivers and associated pressures on aquatic ecosystems is a key barrier to attaining internationally agreed goals directed at addressing their root causes. Lack of appropriate indicators or targets for many environmental, socio-economic and governance goals makes assessing progress towards achieving water-related goals and sustainable aquatic ecosystems especially problematic. Other major barriers include inadequate capacity, limited access to technology and funding, information and data gaps, and lack of quantifable targets. More emphasis, including enhanced monitoring efforts, should be directed to acquiring reliable data on the impacts of climate change and extreme weather events on human health and well-being, and on environmental integrity. Unfortunately, monitoring of water quality, quantity and ecosystem health has been reduced in many regions. As a result, there are increasing uncertainties regarding assessment and management of the water environment, due both to data gaps and to the rapidly changing nature of water issues, including those related to climate change. Data collection, including of gender-specifc data, also requires greater attention, especially regarding the impacts of extreme weather events – storms, foods and droughts – on affeed people. This should form a basis for future policy development, adoption and implementation to enhance the security and livelihoods of all affeed by such events, including women, children and the elderly.

References

Ali, M.H. (2010). *Fundamentals of Irrigation and On-Farm Water Management* Volume 1, and *Practices of Irrigation and On-Farm Water Management* Volume 2. Springer Science+Business Media, New York, NY.

Bates, B.C., Kundzewicz, Z.W., Wu, S. and Palutikof, J.P (eds.) (2008). *Climate Change and Water.* Technical paper of Intergovernmental Panel on Climate Change. IPCC Secretariat, Geneva.

Glennie, P., Lloyd, G.J. and Larsen, H. (2010). *The Water-Energy Nexus: The Water Demands of Renewable and Non-Renewable Electricity Sources.* DHI, Horsholm.

Hoekstra, A.Y. and Mekonnen, M.M. (2011). *Global Water Scarcity: Monthly Blue Water Footprint Compared to Blue Water Availability for the World's Major River Basins.* Value of Water Research Report Series No.53. UNESCO-IHE, Delft.

Pereira, L.A.S., Cordery, I. and Iacovides, I. (2009). *Coping with Water Scarcity: Addressing the Challenges.* Springer Science.

Perret, S., Stefano, F. and Rashid, H. (eds.) (2006). *Water Governance for Sustainable Development: Approaches and Lessons from Developing and Transitional Countries.* Earthscan, London.

5

Governance of Biodiversity

Biodiversity is formally defned by the Convention on Biological Diversity (CBD) as: "the variability among living organisms from all sources including, among others, terrestrial, marine and other aquatic ecosystems and the ecological complexes of which they are part; this includes diversity within species, between species and of ecosystems".

In recent years the links between biodiversity and ecosystem services and the benefits people derive from these have received increasing attention. There is growing evidence that biodiversity has a vital role in attaining the Millennium Development Goals: it contributes to poverty reduction and to sustaining human livelihoods and well-being through, for example, underpinning food security and human health, providing clean air and water, and supporting economic development. Given the importance of biodiversity and evidence of its ongoing decline, it is essential to chart progress in reducing and, as far as possible, reversing the rate of decline.

Recent assessments of the status of biodiversity have shown little evidence of improvement. The third *Global Biodiversity Outlook* (*GBO-3*), was launched in May 2010, and showed that biodiversity has continued to decline since publication of the Millennium Ecosystem Assessment and the last *Global Environment Outlook* (*GEO-4*). The three objectives of the CBD, namely, the conservation of biological diversity, the sustainable use of its components, and fair and equitable sharing of the benefits arising out of the utilization of genetic resources, as well as the missions and objectives of other biodiversity-related conventions are all considered.

It globally agreed indicators and goals for biodiversity, in particular the Aichi Biodiversity Targets. The implications for human well-being of not achieving these targets are examined and gaps in achieving internationally agreed goals for biodiversity are identified, so as to frame key messages for the international community. Current knowledge of the pressures, state and trends affeing biodiversity and of the benefits of biodiversity to people is synthesized from past assessments and recent publications. Management responses that address these pressures are also examined so as to chart progress in safeguarding biodiversity. In particular, cross-boundary issues are tackled from both an ecological and an equity perspective. The links between biodiversity and traditional knowledge and cultural diversity are also considered before concluding with a look to the future.

Goals and targets are one aspect of the policy agenda for assessing progress in meeting global commitments for biodiversity. Eighteen goals related to biodiversity have been identified. These range from the Millennium Development Goal 7 to ensure environmental sustainability, to the most recent five strategic goals and 20 Aichi Biodiversity Targets of the Strategic Plan for Biodiversity 2011–2020. These biodiversity goals and targets have been clustered into themes and prioritized by taking into account the links between them and by reference to key biodiversity issues.

Biodiversity is affeed by multiple drivers and pressures that modify its ability to provide ecosystem services to people. The interaction of multiple drivers, including demographic, economic, socio-political, scientifc and technological ones, is known to increase pressures on biodiversity, leading to further decline, degradation and loss. However, the mechanisms associated with such loss require further research.

Pressures

The principal pressures on biodiversity include habitat loss and degradation, overexploitation, alien invasive species, climate change and pollution (Baillie *et al.* 2010; Vié *et al.* 2009; MA 2005a). These pressures are continuing to increase.

Habitat Loss

Habitat loss in the terrestrial domain has been caused largely by the expansion of agriculture: more than 30 per cent of land has been converted for agricultural production. Large-scale commercial agriculture has adversely affeed biodiversity, particularly agro-biodiversity. Moreover, the growing demand for biofuels has taken a toll, with expanses of forests and natural lands in South East Asia being converted into mono-crop plantations.

Overexploitation

Overexploitation of wild species to meet consumer demand threatens biodiversity, with unregulated overconsumption contributing to declines in terrestrial, marine and freshwater ecosystems. Although overexploitation is often difficult to quantify in terrestrial systems, major exploited groups include plants for timber, food and medicine; mammals for wild meat and recreational hunting; birds for food and the pet trade; and amphibians for traditional medicine and food. The threat to vertebrates from overexploitation is particularly severe, driven, in particular, by demand for wildlife and wildlife products from East Asia. Globally, utilized vertebrate populations have declined by 15 per cent since 1970 as indicated by the Living Planet Index. Similarly, the extinction risk of utilized bird species increased during 1988-2008, driven in part by overexploitation.

In the marine realm, capture fisheries more than quadrupled their catch from the early 1950s to the mid-1990s. Since then, catches have stabilized or diminished, despite increased fishing effort. The proportion of marine fish stocks that are overexploited, depleted or recovering from depletion rose from 10 per cent in 1974 to 32 per cent in 2008. Of the 133 local, regional and global extinctions of marine species documented worldwide over the last 200 years, 55 per cent were caused by overexploitation, while the remainder were driven by habitat loss and other threats. Commercial fisheries are the principal threat to fish stocks, but overexploitation in artisanal fisheries also occurs. Such practices can ultimately lead to major shifts in community composition. For example, coral communities have been transformed into algal-dominated systems because of overfishing of herbivores.

The use of destructive fishing practices further amplifes the impacts of unsustainable fishing on marine biodiversity and habitats. Technology can enhance the intensity and range of human impacts on marine biodiversity although it can also play a significant role in making fishing practices less destructive. Moreover, abandoned and lost fishing gear is having negative ecological consequences on marine biodiversity (also known as ghost fishing).

Overfishing is also a problem in freshwater wetlands, although in many cases adequate data are not available to quantify the extent of the loss. Recreational fishery practices such as stocking and selective take can also have important evolutionary impacts on freshwater fish stocks. By-catch from fisheries can be a major threat to groups such as sharks, turtles and albatrosses.

Invasive Alien Species

Invasive alien species threaten native biodiversity and are spreading through both deliberate and unintentional introductions as a consequence of increasing levels of global travel and trade. Poorly planned economic introductions, air transport, hull-fouling and ballast water from ships, as well as trade in pets, garden plants and aquarium species, are significant pathways for the dispersal of invasive species. Invasive alien species affect native species principally through predation, competition and habitat modification. Invasive species have major economic costs, estimated by one study to total US$1.4 trillion annually. They are found in nearly all countries and habitats, including marine ecosystems – for example the red lionfish *Pterois volitans* affes coral reef fish in the Caribbean – and freshwater ecosystems: the Nile Perch *Lates niloticus*, for instance, has an impact on native fish in Lake Victoria. Invasive species have particularly acute effes on the terrestrial biodiversity of small islands. Data from Europe show that the number of alien species has increased by 76 per cent since 1970, a pattern that is likely to be similar in other places. In another study, invasive alien species were a factor in more than 50 per cent of vertebrate extinctions where the cause was known, and were the sole cause of 20 per cent of extinctions.

Climate Change

Climate change is an increasingly important threat to species and natural habitats. There is widespread evidence that changes in phenology, including the timing of reproduction and migration, physiology, behaviour, morphology, population density and distributions of many different types of species are driven by climate change. For example, trends in European bird populations since 1990 show a growing impact: populations are increasing among the species projected to benefit from climate change while population decline is documented for those projected to undergo range contraction. In the Arctic, tundra habitats are shrinking owing to tree-line advance. In the marine realm, climate change is causing widespread die-of of coral reefs through rising temperatures and ocean acidification. The Arctic ice cap is also shrinking rapidly, with likely impacts on ice-dependent species, as well as shifts in phenology and distribution of marine species (Dulvy *et al.* 2008; Hiddink and Ter Hofstede 2008; Richardson 2008; Perry *et al.* 2005). Recent studies have also projected distribution shifts of 1 066 marine fish and invertebrate species polewards at an average rate of 40 km per decade, leading to likely disruption of community composition and local extinctions.

For many wetlands, changes in rainfall and evaporation are expected to have major impacts on water regimes, affeing both migratory and residential species, while changes in fow in both the short and long term will impact many aquatic species. Climate change will also act synergistically with other threats, such as the spread of diseases and invasive alien species. However, in many instances it may be difficult to differentiate the effes of these different threats, as has been outlined for wetlands and rivers in Australia.

Pollution

Pollutants such as pesticide and fertilizer efuents from agriculture and forestry, industry including mining and oil or gas extraction, sewage plants, run-of from urban and suburban areas, and oil spills, harm biodiversity directly through mortality and reduced reproductive success, and also indirectly through habitat degradation. Inland wetlands and coastal marine habitats face a major threat from waterborne pollutants. Meanwhile, atmospheric pollution in terrestrial systems, particularly the deposition of eutrophying and acidifying compounds such as nitrogen and sulphur, is also important. Rates of nitrogen deposition increased sharply after 1940 but have levelled out since 1990, probably owing to an overall decrease in biomass burning, though there is regional variation. Nevertheless, nitrogen deposition continues to be a significant threat to biodiversity, especially for species that have adapted to low-nitrogen habitats.

Additional Threats

Additional threats to biodiversity include changes in fire regimes, problematic native species and negative influences from human activities. Influences from human activities that may be harmful to biodiversity include artifcial illumination, genetically modified organisms, microplastics, nanotechnology, geo-engineering, and high levels of human appropriation of net primary productivity (Cole 2011; Gough 2011; Galgani *et al.* 2010; Hölker *et al.* 2010; Sutherland *et al.* 2009, 2008). Scientifc understanding of the specifc nature of threats to biodiversity from these influences is building. Meanwhile, the causes of some recent biodiversity declines remain unclear, and further research is required to elucidate the problems and identify solutions, for example for mammals in northern Australia or trans-Saharan migrant birds.

Patterns of Biodiversity Change

Biodiversity is deteriorating at the level of populations, species and ecosystems, and genetic diversity is also suspected to be declining, although trends remain largely unknown. Populations of vertebrate species recorded in the Living Planet Index have declined on average by 30 per cent since 1970. Declines in freshwater populations are steeper, at 35 per cent since 1970, than those for terrestrial populations, which have fallen by 25 per cent and marine populations by 24 per cent; those in the tropics are steeper than those in temperate latitudes. Habitat-specific trends are available for some regions for birds and show, for example, that European farmland bird populations have declined by 48 per cent on average since 1980. North American grassland and dryland species have declined by 28 per cent and 27 per cent respectively since 1968; but North American wetland bird species have increased by 40 per cent.

At the species level, the proportion of species threatened with extinction – classified as critically endangered, endangered or vulnerable on the IUCN Red List – ranges from 13 per cent for birds to 63 per cent for cycads, and averaging almost 20 per cent for vertebrates. Furthermore, Red List Indices for mammals, birds, amphibians and corals show that considerably more species have become more threatened with extinction over recent decades than have become less threatened, and declines have been steepest for corals. The composition of biological communities is increasingly disrupted by human activities, in particular through overexploitation. For example, in some oceans the community structure appears to have shifted to lower trophic levels owing to fisheries targeting predators and larger fish species. This phenomenon of fishing down the food web has been reported widely in many parts of the ocean, such as in Canada, Brazil, India, Thailand, the North Sea and the Caribbean. However, the use of catch data to indicate fishing down the food web may be confounded by data quality and factors such as the spatial expansion of fisheries, and may need careful interpretation if independent data on stock levels are unavailable. Other indicators, such as the Fishing-In-Balance (FIB) index, may be preferable in future.

At the level of habitats, losses include more than 100 million hectares of forest globally during 2000–2005, or 3 per cent of the 3.2 billion hectares in existence in 2000; 20 per cent of mangroves since 1980; and 20 per cent of seagrasses since 1970. Remaining habitats are increasingly degraded – measures of net primary productivity, for example, show that around one-quarter of the terrestrial land area is degraded, including around 30 per cent of all forests, 20 per cent of cultivated zones and 10 per cent of grasslands. Similarly, coral reefs have declined globally by 38 per cent since 1980. Natural habitats are also becoming increasingly fragmented – 80 per cent of remaining forest fragments in the Brazilian Atlantic Forest are now smaller than 50 hectares, while two-thirds of the world's largest rivers are now moderately to severely fragmented by dams and reservoirs.

Benefits to People From Biodiversity

Biodiversity underpins the ecosystem services that supply benefits to people. The deterioration or loss of biodiversity and ecosystem services tends to affect poor people most directly as they are the most dependent on local ecosystems and often live in places that are most vulnerable to ecosystem change. Because the precise mechanisms of human dependence on biodiversity are not fully understood and biodiversity is undervalued – especially for regulating services – maintenance of biodiversity is rarely fully integrated into policy. Progress has been made since the Millennium Ecosystem Assessment, which strongly supported the concept of ecosystem services and their role in providing supporting, provisioning, regulating and cultural services. More recently, The Economics of Ecosystem Services and Biodiversity and green economy approaches have quantified the value of biodiversity and ecosystem services. The Ramsar Convention on Wetlands has further outlined the direct links that exist between ecosystem services from wetlands and human health.

Biodiversity and Human Well-being

Biodiversity and ecosystem services provide food, medicines, fish and timber products as well as biomass, energy and water-related services that people need for their livelihoods and well-being. Too often, the use and management of these provisioning services has failed to focus on conserving the ecosystems providing them. This has resulted in the degradation of regulating and supporting services that are important for overall system functioning and long-term resilience to change and therefore to human well-being, a point that has been well demonstrated when considering the effes of expanding agriculture and water management. Decreases in provisioning services may be a defnitive signal that a biophysical threshold has already been passed with respect to an ecosystem's ability to provide a service, as in the case of a number of fishery collapses.

Food and medicines produced from terrestrial and aquatic ecosystems include wild-harvested products, as well as farmed crops, livestock, fish and aquaculture products. Wild-harvested foods, such as wild meat, non-timber forest products, wild fruits and freshwater resources, remain important for food security, health, cultural identity and adaptation for many people. Likewise, in some Asian and African countries, up to 80 per cent of the population depend on traditional medicines. Assessment of the status of birds and mammals used for these purposes indicates that they are on average facing a greater extinction risk than other species. Although global data for plants are not available, medicinal plants face a high risk of extinction in those parts of the world where people are most dependent on them. This emphasizes the threat posed by biodiversity loss to the health and well-being of people directly dependent on the availability of wild species.

Fisheries provide a major source of food, revenues and employment with globally over 80 million tonnes of biomass being captured annually from the ocean and large amounts from inland waters. However, as fish stocks are depleted, this supply is becoming increasingly dependent on aquaculture, which itself can have many negative environmental and social impacts such as pollution, introduction of exotic species and displacement of small-scale fishing practices. Recent estimates suggest that in 2000 alone, the potential global catch losses due to overfishing amounted to 7–36 per cent of the actual tonnage landed that year, resulting in a landed value loss of US$6.4–36.0 billion. This amount could have helped prevent around 20 million people worldwide sufering from undernourishment.

Agricultural production is also supported by biodiversity and ecosystem services, and agricultural diversity can in turn contribute to food security by supporting adaptation to a changing climate. Small-scale livestock husbandry and pastoralism can both contribute to maintaining biodiversity and to sustainable local economies, adaptation to climate change, resistance to disease and cultural diversity. Equally, overgrazing can cause soil erosion and desertification, thereby decreasing provisioning services. Threats from livestock production to biodiversity are likely to grow as demand for meat and dairy increases, requiring more livestock feed and more water. The complex issue of ensuring a sustainable food supply for an expanding human population has been addressed in recent assessments, along with the biodiversity benefits that can be obtained by balancing food production with the supply of other ecosystem services. Pressures on land, water and biodiversity from agriculture and aquaculture could be reduced in some countries by reducing overconsumption of food, shifting towards diets comprising less meat/fish, and reducing crop losses and food waste.

Energy for much of the world's population is derived from biomass. The most commonly used fuels for heating and cooking are wood, charcoal and plant and animal waste. Hydropower depends on high volumes and regular rates of fow of water to dams from functioning ecosystems in the catchment area, but often contributes to widespread negative environmental and social impacts, particularly loss of biodiversity and displacement. The degradation or loss of ecosystem services that provide energy is evident in the siltation of reservoirs and loss of water volume associated with deteriorating catchment areas; in the deforestation created by the overharvesting of woody vegetation; and in the overuse of agricultural waste and animal manure. The loss of ecosystem services associated with overharvesting, poor management, climate change and, for example, an increase in forest fres, is often felt by already marginalized groups who have to collect fuelwood and/or other forms of biomass for household energy needs. The development of renewable energy from marine and coastal environments, such as that from offishore wind farms, may result in trade-offs between energy production and habitat loss.

Freshwater from surface and groundwater ecosystems is a critical provisioning service used for drinking, sanitation, cooking and agriculture. Wetlands and rivers regulate fows and material cycles that play indispensible roles in supporting human life systems and benefiting many sectors of society. These ecosystems also provide important regulatory services in the form of water purification, erosion control and storm bufering. Meanwhile, groundwater ecosystems provide great social and economic benefits through the provision of low-cost, high-quality water supplies for both urban and rural areas. Groundwater is also important for irrigation, with Siebert *et al.* reporting that 40 per cent of irrigated areas, some 300 million hectares representing about 20 per cent of total farmland, are supplied by groundwater.

Cultural and spiritual values from biodiversity are important to many communities. Many have benefited from exploiting the recreational and cultural value of biodiversity for ecotourism. For example, lakes, wetlands, rivers and coastal ecosystems ofer significant ecotourism potential with, for example, coral reef tourism in Belize estimated to be worth US$150–196 million per year. These aquatic ecosystems also supply water that is integral to many social, spiritual and religious activities. Examples include the sacred status of water sources and riparian zones for the Bantu-speaking peoples of Southern Africa and the duty of care exercised by Maori in New Zealand for the life force exhibited by water.

Biodiversity and Climate Change

Biodiversity plays an important role both in supporting efforts to mitigate climate change and in enabling societal adaptation to its effes. Ecosystems store and sequester carbon through biological and biophysical processes that are underpinned by biodiversity. About 2 500 gigatonnes (billion tonnes) of carbon are stored in terrestrial ecosystems, compared to approximately 750 gigatonnes in the atmosphere. Almost 38 000 gigatonnes are stored in the oceans, of which about 37 000 gigatonnes are in deep ocean layers that will only feed back to atmospheric processes over very long time scales. Around 1 150 gigatonnes are stored in forests, with 30–40 per cent in biomass and 60–70 per cent in soil. Significant carbon stocks are also found in other terrestrial ecosystems including wetlands and peatlands. Indeed, the latter cover only 3 per cent of the land area, but reputedly contain nearly 30 per cent of all global soil carbon. Marine ecosystems on average take up an additional 2.2 gigatonnes of carbon per year. The critical role of freshwaters in the global carbon cycle has only recently been demonstrated.

The importance of forests in storing almost half of all terrestrial carbon, and sequestering carbon from the atmosphere, means that they play a major role in climate change mitigation. Primary forests are more biologically diverse, and also more carbon dense, than other forest ecosystems. Modified natural forests and plantations have less biodiversity and lower carbon stocks than primary forests under similar environmental conditions. Eforts to maintain forest health, for example through incentive mechanisms such as Reducing Emissions from Deforestation and Degradation (REDD+), have the potential to help mitigate climate change. These can also have multiple benefits for biodiversity if interventions ensure that environmental and social safeguards are respected, such as full and effective participation of indigenous and local communities, and if they avoid displacing deforestation and degradation from areas of lower conservation value to those of higher biodiversity value, or exerting pressures on other native ecosystems.

Many of the options available to help society adapt to the effes of climate change depend on and are enhanced by biodiversity. Ecosystem-based adaptation uses the range of opportunities for the sustainable management, conservation and restoration of ecosystems to provide services that enable people to adapt to the impacts of climate change. For example, intact, well-functioning ecosystems, with natural levels of species diversity, are usually more able to continue to provide ecosystem services, and resist and recover more readily from extreme weather events than degraded, impoverished ones. Healthy ecosystems also play an important role in protecting infrastructure and enhancing human security, and therefore in reducing the risk from disasters. Ecosystem-based adaptation options are often more accessible to the rural poor than interventions based on infrastructure and engineering, and there can be multiple social, economic and environmental co-benefits for local communities from its use when designed and managed appropriately.

Responses to The Threats to Biodiversity

Managing Agriculture and Biodiversity

Successful management of agricultural landscapes requires a reduction of habitat loss and degradation whilst providing an adequate supply of food for a growing human population. Sustainable agriculture has received increasing attention because expanding agriculture is globally the principal driver of biodiversity decline. In recent years attention has been given to a new paradigm of ecoagriculture or integrated conservation-agriculture, which seeks to integrate biodiversity conservation with rural development. This paradigm is being explicitly considered in shaping conservation strategies with clearly identified economic and ecological relationships that include ecosystem services. The extensification of agriculture may require more land than intensive agriculture to achieve the same production levels, but it may be more sustainable in the long term and have fewer impacts on wildlife and human health. New approaches that combine the most effective, least harmful practices from intensive and extensive farming, sometimes termed sustainable intensification, will be needed. In this context the use of GMOs in agriculture and also in aquaculture potentially presents both threats and opportunities for biodiversity.

Managing Invasive Species

Successful management of invasive species relies on preventing the introduction and spread of species to new areas, as well as controlling and eradicating established invaders. Ten different international agreements and organizations have some relevance, including the International Plant Protection Convention, the World Trade Organization, the International Maritime Organization, the Convention on International Civil Aviation and the Convention on Biological Diversity. Since 1970 there has been a significant increase in the number of parties to these agreements, with 81 per cent of the world's countries acceding to them. Although this signifes an international intent to manage biological invasions, no international agreement currently deals exclusively with the trade, transport or control of alien and invasive species.

At a national level, only 55 per cent of countries have legislation to prevent the introduction of new ones and to control existing ones, and less than 20 per cent are estimated to have comprehensive strategies and management plans. In many cases, information on existing management activities either does not exist or is not readily available.

To control threats from invasive alien species, the following actions are seen as necessary:

- integrated planning to prevent further introductions by managing priority pathways;
- focus on controlling established species and priority invaders with significant impacts on biodiversity; and
- investment in the knowledge generation, data collation and research needed for risk assessments.

Managing Wildlife Trade and Use

Wildlife use and trade can be managed through a variety of measures, including regulatory measures such as policies and laws and voluntary ones such as certification schemes; formal measures such as positive and economic incentives, and informal ones such as influencing sustainable consumer behaviour; direct measures such as customs inspections and other enforcement actions, and indirect ones such as economic influences. These measures can be applied at a variety of levels from the local, such as delineating resource extraction zones in protected areas or establishing community-based natural resource management, to the global, such as through the Convention on International Trade in Endangered Species of Wild Fauna and Flora (CITES).

Managing the Impacts of Climate Change on Biodiversity Through Mitigation and Adaptation

Managing the impacts of climate change will be important, as recent studies show that the range shifts of terrestrial organisms towards the poles and higher altitudes as a result of climate change may be significantly faster than previously thought. Minimizing the adverse impacts of climate change on biodiversity is dependent on:

- efforts to mitigate climate change itself; measures to ensure that those activities and societal adaptation efforts do not themselves have adverse impacts on biodiversity; and
- application of best practice in conserving and restoring biodiversity in the face of climatic change.

Of the wide range of approaches, many are dependent on the conservation and sustainable use of healthy ecosystems, and ofer opportunities for synergies in terms of climate change mitigation and maintenance of biodiversity. In particular, this concerns intact forests and wetlands, but also natural and semi-natural grasslands and many agricultural ecosystems. For example, some agricultural approaches, such as conservation tillage and agroforestry, can result in the maintenance and enhancement of terrestrial carbon stocks and also contribute to the conservation and sustainable use of biodiversity. Traditional knowledge and systems of small-scale livestock husbandry, farming, and forest product collection can greatly enable local mitigation and adaptation in culturally appropriate ways. However, ecosystem-based approaches also carry risks and these need of intact and functioning ecosystems supported by restoration initiatives wherever possible.

Managing Area-based Conservation

Protected areas are seen by many as the core means of preventing ongoing losses of species and habitats. Protected areas have expanded over the past 20 years in both number and area and now cover 13 per cent of the world's land area. However, coverage is uneven, and 6 of the 14 global biomes and half of the 821 terrestrial ecoregions do not meet the CBD target for 10 per cent of their area to be protected by 2010. Furthermore, the expansion of the world's protected area network needs to be targeted at the most important sites for biodiversity. Some 51 per cent of the 587 sites identified by the Alliance for Zero Extinction as critical for the survival of hundreds of highly threatened species, and 49 per cent of the more than 10 000 important bird areas are still entirely outside the protected area network. Even more importantly, the performance of protected areas in maintaining populations of their key species is poorly documented. Although some studies have found wildlife declines within some protected areas, others demonstrate that protected areas have been effective in maintaining species that would otherwise have disappeared. However, not all species may require protected areas to ensure their survival, and protected areas require complementary broad-scale conservation measures.

Uneven protected area coverage of biomes is most evident in the marine realm, despite a CBD target to protect 10 per cent of the ocean by 2012. By the end of 2010 marine protected to be assessed and addressed. In the case of forests, the United Nations Framework Convention on Climate Change and CBD have recognized a need for safeguards, particularly for biodiversity and human communities, to minimize risks associated with REDD+. There is also a risk of confict between the goals of the Kyoto Protocol's Clean Development Mechanism for carbon sequestration and biodiversity conservation.

Other concerns about the impacts of mitigation activities include those related to artifcial ocean fertilization through using nutrients such as iron or nitrogen to increase the uptake and sequestration of atmospheric carbon. The effectiveness of this approach is highly uncertain and increasingly thought to be quite limited. Potential negative environmental effes include increased production of methane and nitrous oxide and changes in phytoplankton community composition, which may lead to toxic algal blooms. Alternative sources of energy production, such as biofuels, hydropower, wind farms and oceanic tide generators have all been documented to have impacts on biodiversity if safeguards are not developed. The most fundamental strategy for conserving biodiversity in the face of climate change will continue to be promoting the conservation areas covered 1.6 per cent of the ocean area. Indeed, at the end of 2010, only 12 countries had designated more than 10 per cent of their waters, often through large areas, while 121 countries had yet to designate more than 0.5 per cent of their marine jurisdiction. In response, the CBD has retained the 10 per cent target, with a revised achievement date of 2020.

Marine protected areas can be designated at a variety of levels of protection, but those with complete protection provide the greatest biodiversity benefits. A review of 112 independent studies in 80 different protected areas found significantly higher fish populations inside the reserves than in surrounding areas or in the same place before protection was established. Relative to reference sites, population densities were 91 per cent higher, biomass 192 per cent higher and average organism size and diversity 20–30 per cent higher, usually between one and three years after establishment of a reserve. These trends occurred even in small marine protected areas.

Protected areas can also play a key role in climate change mitigation and adaptation, preventing the conversion of natural habitats to other land uses and hence avoiding significant release of carbon. Emissions from land-use change, mainly forest loss, contribute up to 17 per cent of all anthropogenic greenhouse gas emissions. It has been estimated that about 15 per cent of the global terrestrial carbon stock is stored in the world's protected area network, and the role that this can play in climate mitigation is underlined by the fact that between 2000 and 2005, protected areas in humid tropical forests lost about half as much carbon as the same area of unprotected forest.

Indigenous and Community-conserved Areas

Protected areas can be effectively managed by many groups, from government agencies to local communities and indigenous peoples, and from non-governmental organizations (NGOs) to private individuals. Recently, the full range of IUCN protected area categories has been brought into use for designating protected areas. For example, in Australia, protected areas established and managed by indigenous communities comprise nearly a quarter of Australia's national reserve system by area. Indigenous and community-conserved areas (ICCAs) and sacred natural sites (SNSs) have proven successful in conserving a rich biological and biocultural diversity by supporting the maintenance of traditional environmental knowledge and practices. These community areas are extremely diverse, manifesting myriad ethical, economic, cultural, spiritual and political dimensions (Brown and Kothari 2011; Borrini-Feyerabend *et al.* 2010a, 2010b; Kothari 2006; Posey 1999). They include waterfowl nesting wetlands, roosting sites or other critical wild animal habitats, and landscapes with mosaics of natural and agricultural ecosystems such as the Potato Park in the Andean Highlands of Peru and the rice terraces of the Philippines. A number of studies demonstrate the wide range of values they provide.

The number and extent of ICCAs and SNSs have not been comprehensively estimated. It has been, nevertheless, suggested that in some parts of the world their area is similar to that currently under government-managed protection. Furthermore, it has been estimated that communities own or manage 22 per cent of all forests in 18 developing countries. Recent analyses highlight the potential effectiveness of indigenous and community-managed areas in tropical forest conservation. For example, such areas can be more effective in reducing tropical deforestation than forest protected areas, and indigenous and multiple-use protected areas can reduce the incidence of tropical forest fres as effectively as strict protection.

ICCAs and SNSs are increasingly recognized as legitimate and powerful tools for the security of both their custodians and the biodiversity they encompass, supported by a range of conservation, human-rights and development instruments. A preliminary survey of the laws and policies of 27 countries and one sub-national region showed that progress in national recognition of ICCAs and SNSs is patchy: some countries are moving rapidly, others slowly, and some not at all. The biggest challenge, now that ICCAs and SNSs have global attention, is in gaining appropriate national recognition and support, particularly for tenure, customary practices and decision-making institutions, and other fundamental human rights. Activities relating to governance, participation, equity and benefit sharing in relation to protected areas merit increasing consideration.

Recognizing the Value of Cultural Diversity and Traditional Knowledge

The recognition of human and natural systems as unified social-ecological systems is increasingly important for safeguarding biodiversity. This growing understanding underscores the links between biological and cultural diversity and the role of local and indigenous peoples in the sustainable governance and management of biodiversity. The Strategic Plan for Biodiversity and the Aichi Biodiversity Targets support greater respect of traditional knowledge and its full integration and reffeion in CBD implementation at all levels, with the full and effective participation of indigenous and local communities. Information on the status and trends of linguistic diversity has been used as a proxy indicator for traditional knowledge, innovations and practices, including those about biodiversity. Traditional knowledge is an invaluable and irreplaceable source of information about biodiversity and human relationships; its loss entails a loss of collective cultural heritage and capacity to adapt to and live sustainably within specifc ecosystems and areas.

Access and Benefit Sharing of Genetic Resources

The fair and equitable sharing of the benefits of exploiting genetic resources is one of the three CBD objectives (Article 1), recognized as critical for biodiversity conservation. Through the recently adopted Nagoya Protocol on Access to Genetic Resources and the Fair and Equitable Sharing of Benefits Arising from their Utilization, standards are established for regulating access to genetic resources and the distribution of benefits from their use, as well as the associated traditional knowledge. The principle underlying CBD recognizes that states have a sovereign right to exploit their own resources pursuant to their own environmental policies (Article 3).

Access to genetic resources has emerged as a major political rallying point in international negotiations. Much of the world's biodiversity is concentrated in the forests of developing countries in the tropics, but much of the technology and financial capital that can convert elements of biodiversity into commercial products rests with the developed countries. Hence, while unprecedented biodiversity loss is a global concern, commercial use and the associated issues of intellectual property fundamentally alter the nature of biodiversity as a global public good. The impetus behind the Nagoya Protocol arose from growing discontent amongst developing countries and indigenous and local communities regarding the lack of implementation of the benefit-sharing provisions of CBD since it came into force in 1993. This was compounded by only a handful of user countries undertaking any compliance measures to prevent bio-piracy despite the adoption of guidelines in 2002.

The Nagoya Protocol is an important milestone for rectifying the issues of equity associated with the commercial use of genetic resources and associated traditional knowledge. The protocol is also unprecedented in its recognition of the right of indigenous and local communities to regulate access to traditional knowledge associated with genetic resources in accordance with their customary laws and procedures. The protocol opened for signature in February 2011 and will not enter into force until 90 days after 50 countries have signed. A number of countries already have national legislation and regulations pertaining to issues of access and benefit sharing, and monitoring the further development of such regulations could provide a useful indicator of progress.

In the marine realm, ten countries own 90 per cent of the patents deposited with marine genes – with 70 per cent belonging to just three – but account for only about 20 per cent of the world's coastline. These nations benefit from access to the advanced technologies required to explore the vast genetic reservoir of the oceans, leading to a call for policies targeting capacity building to improve access for other countries.

Assessment of Progress and Gaps Conservation Strategies

Protected areas are one of the primary responses for maintaining biodiversity, particularly on land, but are generally deemed to be insufcient. The exclusion of local communities from many state and private protected areas along with the failure to fully acknowledge their role in safeguarding biodiversity remains a challenge to real progress. Outside protected areas the proportion of sustainably managed production landscapes – for agriculture, forestry, fisheries and aquaculture, amongst others – is increasing, but only slowly. For example, the area of forest certified by the Forest Stewardship Council (FSC) as sustainably managed continues to grow, reaching 149 million hectares in 2012, and there are additional areas managed under the Programme for the Endorsement of Forest Certification (PEFC), but this remains a small fraction of the global total of managed forests. Similarly, fish products certified by the Marine Stewardship Council (MSC) constituted only 7 per cent of global fisheries in 2007.

National Biodiversity Strategies and Action Plans

The CBD requires all member states to develop a national biodiversity strategy and action plan as the primary mechanism for the implementation of its strategic plan. To date, 172 of the 193 signatory countries have adopted their plans or equivalent instruments. The large number of plans is an achievement in itself, and more so where they have stimulated conservation action at the national level and contributed to a better understanding of biodiversity, its value and management. In spite of these achievements, national strategies have not been fully effective in addressing the main drivers of biodiversity loss. Only a few countries have used the plans as mechanisms for mainstreaming biodiversity and ecosystem services, and there is generally poor coordination with other relevant policies. However, Parties to the CBD are expected by 2014 to revise their plans in line with the new Strategic Plan for Biodiversity 2011–2020, which includes reference to improving mainstreaming.

Resource Mobilization

Many national reports submitted to the CBD have identified the lack of financial, human and technical resources as the most widespread obstacle to implementation of national strategies and the CBD in general. Thus, the fulfilment of the Aichi Target to substantially increase resource mobilization will be crucial to enable the other targets to be achieved.

While documentation is lacking for both the current and the required level of fnancing to safeguard biodiversity, there is no doubt that the gap between the two is substantial. Estimates suggest that existing fnancing is in the order of tens of billions of dollars a year, while total needs are of the order of hundreds of billions a year. International fnancing for biodiversity is estimated to have grown by approximately 38 per cent in real terms since 1992 and now stands at US$3.1 billion annually. The Global Environmental Facility (GEF) will provide US$1.2 billion for CBD implementation from 2010 to 2014, an increase of 29 per cent compared to the previous four years.

Increasingly, innovative financial mechanisms are considered essential tools to mobilize additional resources for biodiversity. These include payment for ecosystem services, biodiversity offsets, ecological fscal reforms, markets for green products and biodiversity in new sources of international development fnance.

Knowledge Gaps for Biodiversity Monitoring

Although indicators of the state of biodiversity are predominantly showing declines, there are considerable gaps in their geographic, taxonomic and temporal coverage. While biodiversity loss is a global phenomenon, its impact may be greatest in the tropics where available indicators and data coverage are the least complete.

Particular gaps in knowledge for state indicators include: grassland and wetland extent, habitat condition, primary productivity, genetic diversity of wild species, freshwater and terrestrial trophic integrity, ecosystem functioning and ocean acidification. Pressure indicators lack data on pollution, exploitation in terrestrial and freshwater ecosystems, wildlife disease incidence and freshwater extraction. The principal gaps in response indicators include sustainable management of agriculture and freshwater fisheries, and management of invasive alien species.

A prominent gap in knowledge concerns ecosystem services. Indicators of the biodiversity that underpins these services should be tailored to the scales at which ecological processes that produce the services occur, such as the landscape scale for agriculture and biomass production, and the watershed for direct water use and hydroelectricity generation.

Other responses to biodiversity loss include policy action to tackle an array of issues including hunting and pollution, and enforcement of environmental impact assessments and mitigation measures for infrastructure development; however, global trend data are unavailable for these. Given that most global biodiversity targets, such as the Aichi Targets, require action at the national scale, national biodiversity data are crucial for tracking progress towards global biodiversity targets, and to inform national strategies. National Red Lists of threatened species are one of the many examples of nationally relevant biodiversity data that may provide suitable input for reporting on progress towards these goals and for informing national conservation priority setting, although there are others which are also suitable. The Group on Earth Observations Biodiversity Observation Network (GEO BON) is expected to make an important contribution to future monitoring efforts, whilst the Biodiversity Indicators Partnership is supporting global and national biodiversity indicator development for the Aichi Targets and for national biodiversity strategies and action plans.

Projections, Scenarios and Horizon Scans

While recognizing a time frame of increasing uncertainty, synthesizes biodiversity studies from short-term projections through to longer-term scenarios with a view to distilling relatively short-term policy implications. This relies heavily on the *GBO-3* analysis of biodiversity scenarios, for which scientists from a wide range of disciplines came together to seek consensus on projections and scenarios for biodiversity change during the 21st century.

Although quantitative projections and scenario methods are well advanced, the range of projected changes reported by the studies reviewed is rather broad, partially because there are significant opportunities to intervene through better policies, but also because of large uncertainties in the projections. The projections of global change impacts on biodiversity show continuing and, in many cases, accelerating species extinctions, loss of natural habitat, and changes in the distribution and abundance of species and biomes over the 21st century. Possible thresholds, amplifying feedbacks and time-lag effes leading to tipping points appear to be widespread and make the impacts of global change on biodiversity hard to predict, difficult to control once they begin, and slow and expensive to reverse once they have occurred. For many important cases, the degradation of ecosystem services goes hand-in-hand with species extinctions, declining species abundance or widespread shifts in species and biome distributions; however, the conservation of biodiversity and of some services, especially provisioning services, is often at odds. Strong action at international, national and local levels to mitigate the drivers of biodiversity change and to develop adaptive management strategies could significantly reduce or reverse undesirable and dangerous biodiversity transformations if urgently, comprehensively and appropriately applied.

Policy Implications

The accumulated evidence, cited above, suggests that there is greater success in halting the negative changes in biodiversity and ecosystem services when a proactive attitude in support of a sustainable environment is adopted. Overall, the above synthesis, coupled with inputs from UNEP's Foresight Initiative, suggests that:

- land must be used more efciently to decrease the rate of habitat loss;

- mitigation of climate change is urgent and there is a significant risk of tipping points near or even before the 2°C mean global surface temperature target agreed at the UNFCCC meeting in Cancun in 2010;
- payments for ecosystem services and the greening of national accounts can help to protect biodiversity if appropriately applied;
- protected areas by themselves have not been adequate to achieve the target of reducing the rate of biodiversity loss by 2010;
- potential collapse of oceanic ecosystems requires an integrated and ecosystem-based approach to ocean governance; and
- recognizing the importance of local participation and community support, it is crucial to ensure that policies are integrated, sensitive and inclusive of local communities. This applies to conservation strategies, preservation of local cultures and languages, and access and benefit sharing of genetic resources and traditional knowledge.

References

Bates, B., Kundzewica, Z.W., Wu, S. and Palutikof, J. (eds.) (2008). *Climate Change and Water.*

IPCC Technical Paper VI. IPCC Secretariat, Intergovernmental Panel on Climate Change, Geneva.

Emerson, C. (1999). *Aquaculture Impacts on the Environment.* Cambridge Scientific Abstracts.

Finlayson, C.M. and DfCruz, R. (2005). Inland water systems. In *Ecosystems and Human Wellbeing: Current State and Trends: Findings of the Condition and Trends Working Group* (eds.Hassan, R., Scholes, R. and Ash, N.). pp.551583. Island Press, Washington, DC

Molden, D. (ed). (2007). *Water For Food, Water For Life: A Comprehensive Assessment of Water Management in Agriculture.* Earthscan, London and Water Management Institute, Colombo.

6

Chemicals and Waste Management

More than 248 000 chemical products are commercially available and subject to regulatory and inventory systems. Chemicals provide valuable benefits to humanity including in agriculture, medicine, industrial manufacturing, energy extraction and generation, and public health and disease vector control. Chemicals play an important role in achieving developmental and social goals, especially for improving maternal health, reducing child mortality and ensuring food security, and advances in their production and management have increased their safe application. Nonetheless, because of their intrinsic hazardous properties, some pose risks to the environment and human health. Simultaneous exposure to many chemicals – the cocktail or synergistic effect – is likely to exacerbate the impacts.

Chemicals are released at many steps in their life cycle, from the extraction of raw materials, through production chains, transport and consumption, to fnal waste disposal. They are distributed through indoor environments, food and drinking water, and through soils, rivers and lakes. Certain long-lived chemicals such as persistent organic pollutants (POPs) and heavy metals are transported globally, reaching otherwise pristine environments such as rain forests, deep oceans or polar regions, and can quickly pass along the food chain, bioaccumulating to cause toxic effes in humans and wildlife.

Products derived from chemicals often become hazardous wastes in their end-of-life phase, generating additional pollution risks that can devalue their initial benefits and counteract development advantages. Pollution from dumping and uncontrolled open burning is common, and is even increasing in some parts of the world, though some progress has been made in recent decades. The causes of mismanagement often lie in such factors as defciencies in institutional and regulatory frameworks. Such shortcomings also have an impact on the growing transboundary movement of hazardous wastes from developed to developing countries, where compliance, monitoring and enforcement of regulations tend to be weak, and the financial and technical capacity to implement improved waste management practices is limited. This leads to a risk of rapidly increasing exposure for greater portions of the population and to related, often serious, health problems, in particular for women and children.

Broadly, a two-speed situation exists, with developed countries generally having comprehensive systems for chemical and hazardous waste management, while developing countries generally do not. Developing countries and economies in transition struggle with basic landfll co-disposal of many types of wastes, with little capacity for their separation and sound management.

While many developing countries have ratified the multilateral environmental agreements on chemicals and wastes – such as the Basel Convention on the Control of Transboundary Movement of Hazardous Wastes and their Disposal – these are not always transposed into national legislation in a comprehensive manner. In addition, given the cross-sectoral nature of the issue, the regulation and management of chemicals in most developing countries is spread over several ministries – including agriculture, industry, labour, environment and health – and between several agencies within each ministry.

In most countries, it is the poorest members of the population that are at particular risk of exposure. This may be due to occupational exposure, poor living conditions, lack of access to clean water and food, domestic proximity to polluting activities, or a lack of knowledge about the detrimental impacts of chemicals – or a combination of these factors.

Radioactive contamination is another source of potential environmental and health hazards, both from controlled emissions and waste management, and from accidental release. The controlled release of radionuclides to the atmospheric and aquatic environments may occur as authorized efuent discharge, while uncontrolled release may occur as a result of accidents and at legacy sites left by nuclear weapons testing. The management and disposal of radioactive waste from industry, research and medicine, as well as from nuclear power, is relevant to almost all countries, requiring different approaches according to the volume, radioactivity and other properties of the waste.

Initially, governance instruments for chemicals and wastes could be considered to have been reactive, piecemeal and isolated, and with mixed success – the Montreal Protocol on Substances that Deplete the Ozone Layer, for example, being effective in reducing the impact of ozone-depleting substances, while the Basel Convention has struggled to reduce the transboundary movement of hazardous waste. There have been significant advances over the past decade, however, and regulatory instruments are now improving with the better and more widespread understanding of the life cycle of chemicals and their association with the generation and processing of wastes. Eforts to bring the work of the Basel, Rotterdam and Stockholm Conventions together constitute a first step towards addressing the entire life cycle of chemicals. This also applies to the establishment of the Strategic Approach to International Chemicals Management (SAICM) and the current negotiation for an international agreement on mercury. Similarly, the Joint Convention on the Safety of Radioactive Waste Management and the Safety of Spent Nuclear Fuel Management is a significant step forward. However, ensuring that these efforts are sustained and fully anchored at the national level requires further investment in better science-based understanding of chemicals and wastes, policy creativity to balance development and sustainability imperatives, public-private partnerships to link technological innovation and societal responsibility, and allocation of funds for comprehensive capacity building.

Internationally Agreed Goals

It evaluates progress towards internationally agreed goals relevant to chemicals and wastes. The goals are those identified by the *GEO-5* High-Level Intergovernmental Advisory Panel from key multilateral environmental agreements and related agreements and declarations, as further considered and prioritized in regional consultations. The current lack of data, a key constraint on many aspects of chemical and waste management, has not been seen as a reason to preclude the selection of a goal.

In the 1970s and 1980s the human health and environmental impacts of chemicals and waste led to the creation of a number of key international agreements. These, along with other related goal-based international agreements and declarations such as those emanating from the 2002 World Summit on Sustainable Development (WSSD) in Johannesburg, constitute a framework for organizing and implementing specifc goals for the environmentally sound design, production, consumption and recycling or disposal of chemicals and hazardous waste. These goals are also considered against the background of the Millennium Development Goals (MDGs), specifically MDG 1 for eradicating extreme poverty and hunger, and MDG 7 for ensuring environmental sustainability. MDG 7 includes specifc targets for ozone-depleting substances, as well as for improved access to safe drinking water and sanitation facilities.

The broad set of principles pivotal to the development of international agreements comprises prior informed consent for the transboundary movement of hazardous waste and certain hazardous chemicals; transparency through national reporting; the environmentally sound management of chemicals and waste; waste prevention; the precautionary approach; and the polluter-pays principle. These are addressed through specifc obligations such as the implementation of control measures, monitoring of the state of the environment, and compliance regimes with supportive delivery mechanisms including capacity building and training, international cooperation, synergies and partnerships.

Goals relevant to the sound management of chemicals and waste aim to protect human health and the environment while improving resource efciency. They can be grouped into six themes:

- sound management of chemicals throughout their life cycle, including persistent organic pollutants and heavy metals, and of waste;
- control of the transboundary movement of hazardous wastes as well as responsible trade in hazardous chemicals;
- transparent science-based risk assessment and risk management procedures, as well as monitoring systems at the national, regional and global levels;
- support for countries to strengthen their capacity for the sound management of chemicals and waste;
- protection and preservation of the marine environment from all sources of pollution;
- safe radioactive and nuclear waste management.

Data and Indicators

The lack of data on existing chemicals, and the rapid technological changes that bring new chemicals to the market, have hindered the production of an established set of indicators with time-series data that can be used to identify the state and trends of chemicals and wastes. Several possible indicators to fll this gap are proposed below. In addition, extensive investment in collating the required data and solidifying the knowledge base is required to construct long-term time series.

Underlying data on waste generation, treatment and recycling are difficult to obtain. Some are available on hazardous waste through reports to the Secretariat of the Basel Convention, providing information on the quantity, characteristics, destination and mode of treatment or disposal of hazardous waste that is subject to international movement, but even this is incomplete and unverified - as was reported in 2011 to the tenth Conference of the Parties to the Convention. Global data on non-hazardous waste generation and disposal have not been systematically reported and are therefore unsatisfactory. As stated by the UN Secretary-General in his May 2011 report to the Commission on Sustainable Development: *"The barriers to effective management and minimization include lack of data, information, and knowledge on waste scenarios, lack of comprehensive regulations and weak enforcement of existing legislation, weak technical and organizational capacities, poor public awareness and cooperation, and lack of funds."*

There is an urgent need to improve the availability and quality of these basic datasets, with a focus on comparability between countries, timeliness and coherence over time, and interpretability. As waste is increasingly seen as a potential resource, waste data and indicators should be more closely linked to economic and social information systems and material.

Three indicators to help inform governments and municipalities of industry performance and progress are highlighted here. It is imperative that data for these indicators are generated to guide decision making on sound global management of wastes. The key indicators proposed are:

- quantity and types of waste–solid, organic, hazardous and non-hazardous - managed or fnally disposed of;
- waste and hazardous waste generation per person; and
- the amount of municipal or household waste, industrial solid waste and hazardous waste that is recycled.

Status and Trends of The Chemical Industry

The chemical industry is a major driver of economic growth and its performance is a leading indicator of economic development. In 2008 the global chemicals industry had an estimated turnover of about US$3.7 trillion and was growing at 3.5 per cent per year. More than 20 million people around the globe are employed by it directly or indirectly, and it is an intensive energy consumer and a ubiquitous generator of emissions.

While companies in the countries of the Organisation for Economic Co-operation and Development (OECD) continue to account for the bulk of world production (74.5 per cent in 2004), the OECD's share is now 9 per cent less than in 1970. Much of this shift has been caused by the major emerging economies, particularly the BRIC countries (Brazil, Russia, India and China). In 2004, China accounted for most BRIC production (48 per cent), followed by Brazil and India (20 per cent each), and Russia (12 per cent). Chemical consumption in developing countries is likewise growing much faster than in developed countries, and could account for a third of global consumption by 2020. At the same time, some data show that developed countries are reducing chemical use. For example, overall use of pesticides in OECD countries declined by 5 per cent during 1990–2002, although trends vary from country to country. Total releases and transfers of the 152 pesticides that are common to the United States and Canada dropped by 18 per cent and the production of ozone-depleting substances almost stopped; emissions of acid rain precursors dropped by 48 per cent, ozone precursors by 38 per cent and non-methane volatile organic compounds by 26 per cent. Nonetheless, international cooperation between all governments is needed to build capacity, share information and promote effective chemicals management globally.

Waste as an Issue of Global Significance

The growing interdependence of the global economy along with the increasing generation and complexity of waste worldwide can lead countries to unsound waste management and disposal operations, and there may come a point when related costs are such that the economy and public services fail to keep up. Integrated policies are required to support sustainable economic development through recycling, recovery, reuse and other operations aimed at reducing both the use of natural resources and the quantities of waste, as it is inevitable that some resource inputs to industrial production are returned to the environment as waste, and may be hazardous. A critical issue is to reverse current trends in waste generation, which would require a high level of commitment to minimize both quantities and levels of hazard. Furthermore, unsound recycling comes with the risk of pollution and increased human exposure to toxic substances. Recycling can also be misused as a disguise for criminal operations.

The introduction of many new chemical substances to the market leads to the production of new kinds of wastes. In many regions, hazardous waste streams are mixed with municipal or solid wastes and then either dumped or burned in the open. This raises issues of environmental and social justice, as the people most affeed by such precarious practices are usually the poor who live and work adjacent to dump sites.

Through globalization, materials may be produced in one country or region, used in another and managed as waste in a third. Electrical and electronic equipment provides a case in point. The treatment of end-of-life electronics, including toxic substances and plastics with associated fame-retardants as well as precious metals, exemplifes the two sides of this business. The original equipment has the potential to contribute to protecting human health, supporting livelihoods and creating jobs, while also promoting a shift from waste to resources that supports economic development, energy efciency and the conservation of natural resources. However, unsound or inadequate waste management can have profound human health impacts and cause serious harm to the environment. Extending the useful life of electrical and electronic equipment and using less harmful substances in these products is one way to reduce the waste burden and its accompanying hazards.

Municipal Waste

The unsound management of waste can lead to mutually reinforcing undesirable effes. It can pollute and contaminate the environment, pose a threat to human health and represent a loss of resources in the form of both materials and energy. The recent UN-Habitat report on solid waste management in cities refers to the escalating challenge of managing it across the globe, and amply demonstrates the complexity and variety of issues faced, including the difficulty of achieving objectives when progress goes unrecorded, stating for example that "waste reduction is desirable but, typically, it is not monitored anywhere".

Municipal waste constitutes a significant percentage of the total waste a country generates, with annual figures ranging from 0.4 to 0.8 tonnes per person, and solid waste generation increasing at an estimated rate of about 0.5–0.7 per cent per year. Waste complexity is also increasing with the co-disposal of assorted waste types: biodegradable components currently account for almost 50 per cent of municipal solid waste and electronic waste (e-waste) for 5–15 per cent. The management of waste is further complicated by the range and diversity of waste generators, from mining and a wide variety of manufacturers through agricultural and medical waste to household rubbish. In addition, the sound management of municipal waste constitutes a sizable and continuous part of a municipality's budget.

Many countries do not have the infrastructure to deal with ever more complex waste streams. Nor do many have the regulatory and physical infrastructure to derive some rebate from the recyclable materials that are inevitably part of municipal waste.

Life-cycle Thinking

What ultimately determines how humans and ecosystems are exposed to toxic chemicals is defned by their life-cycle characteristics. Releases of substances not only occur during chemical production but also during the use of products containing chemicals, and fnally at their disposal. Life-cycle thinking promotes an integrated approach to the sustainable production and consumption of such substances.

The entire life cycle of resource use, from extraction and production/ manufacture through consumption/use to post-consumption disposal, produces undesirable environmental impacts from emissions and wastes. These impacts can include unintended side effes such as endocrine disruption, which directly interferes with growth and development in most animals, and can also affect people. Life-cycle analysis helps understand such impacts, but, while a useful tool, it can be extremely complex. Too often, when problems are identified, shifts to alternative chemicals that have the same intended properties may result in further unexpected or undesirable outcomes.

The latest materials to raise concern are those arising from synthetic biology and engineered nanomaterials. With the accelerated pace at which new technologies and chemicals are being deployed, a different approach is needed in which their implications are systematically and comprehensively assessed before they reach production. The use of green chemistry principles in chemical design and the adoption of clean production processes may help to prevent problems at a later stage. While this is happening in some parts of the world through the use of exposure models – for example by the Canadian Centre for Environmental Modelling and Chemistry – for some technologies and chemicals, life-cycle analysis has yet to become a universal systematic approach. This may well require new forms of international governance.

The high number and diversity of chemicals and the complexity of their life cycles inevitably lead to a situation where the scientifc understanding of the impacts of chemicals, and the regulatory schemes used to manage them, lag behind technological and economic developments.

Poverty and Exposure to Chemicals

The overwhelming majority of impacts from unsafe chemical use and unsound waste disposal – including death, impairment of health and ecosystem degradation – occur in situations of poverty. Increased risks of exposure to toxic and hazardous chemicals and wastes predominantly affect the poor, who routinely face such risks because of their occupation, poor living standards and lack of knowledge about the detrimental impacts of exposure to these chemicals and wastes. Many of the poor enter the informal sector of the economy where they may encounter new kinds of toxic hazards such as electronic and electrical waste (e-waste). Risk is not only related to the dose they receive from such exposure, but also to important factors such as age, nutritional status and co-exposure to other chemicals. Children are particularly susceptible to the negative health impacts of chemicals due to their rapid growth and development and greater exposure relative to body weight.

A recent study by the World Health Organization (WHO) indicated that 4.9 million deaths were attributable to environmental exposure to chemicals in 2004. Indoor smoke from the use of solid fuels, outdoor air pollution and second-hand smoke are among the most critical causes. The study concluded that the known burden of chemicals, while considerable, is an underestimate because data on many chemicals are scarce.

Changes in the global production, trade and use of chemicals and the concomitant production of hazardous wastes are not always accompanied by corresponding control measures, thus increasing the risk of releasing hazardous chemicals into the environment. It is estimated that there are 2 million contaminated sites in Europe, the United States and the Russian Federation alone. Data for developing countries and economies in transition are more difficult to obtain, but indications are worrying. The Global Inventory Project – which involves the Blacksmith Institute together with the United Nations Industrial Development Organization (UNIDO), the Green Cross and the European Commission – is currently assessing the state of contaminated areas in 80 countries worldwide, with trace metal and pesticide pollution among the ten most problematic types of contamination. This is the first such attempt to give governments, international organizations and affeed communities aggregated data for decision making.

Marine Pollution

The oceans cover 71 per cent of the Earth's surface and are polluted to varying degrees, threatening marine life, fisheries, mangroves, coral reefs, and estuarine and coastal zones, with somc 80 per cent of the pollution coming from land-based sources. Common man-made pollutants include pesticides, chemical fertilizers, heavy metals, detergents, oil, sewage, plastics and other solids. Many of these pollutants accumulate in the deep oceans and sediments, where they are consumed by small marine organisms and may be reintroduced to the global food chain. Some 20 per cent of marine pollution originates from direct disposal into the oceans: regular discharges of oily wastes from ships, accidental oil spills and untreated sewage disposal in enclosed areas such as the Mediterranean are threats to marine ecosystems.

Persistent Organic Pollutants

Persistent organic pollutants (POPs) are a group of chemicals with common features including persistence, bioaccumulation and long-range transport. Combined with their toxicity, these characteristics have significant adverse effes both on wildlife, including marine mammals, and on human populations, in particular such vulnerable groups as nursing mothers and infants. The health effes of exposure to POPs include neuro-developmental disorders, endocrine disruption and carcinogenicity.

The Stockholm Convention on POPs was adopted in 2001 in response to an urgent need for global action, and entered into force in 2004. It currently has 177 Parties and calls for documentation of the amounts of POPs that are still present in different countries and for global monitoring of these substances in human tissue (blood and milk). This is one of two indicators proposed for monitoring and assessing the status and trends of POPs in the environment and their impact on human health. The Stockholm Convention established a Global Monitoring Plan as a source of globally consistent and reliable data. Collection of data is at an early stage and more will become available in the coming years, but individual studies already provide historical and regional trends for some substances. An example is DDT, for which Ritter *et al.* report global time series of concentrations in human tissue from many individual measurements. In general, DDT body burdens have fallen over recent decades, but are still considerably higher in tropical than northern regions. Where DDT is used for malaria control, concentrations are still very high and the decrease is less pronounced than elsewhere.

The other indicator for POPs is trends of selected atmospheric POPs in both urban/industrialized and remote regions. Concentrations of these substances in the air follow changes in emissions more closely than concentrations in food and human tissue, and reffe the effect of atmospheric long-range transport. Hung *et al.* provide a summary of long-term time trends of various POPs measured at Arctic monitoring stations. In general, concentrations of most substances in Arctic air show a falling trend, but half-lives are often long – five to ten years and sometimes even longer. In recent years the decrease has come to a halt for several compounds, and some concentrations have been observed to be on the rise, for example polychlorinated biphenyls (PCBs), chlordane and DDT.

The environmental behaviour of POPs is strongly affeed by temperature and other climate-related factors, including precipitation patterns, wind felds and extreme weather events. In general, climate change is expected to cause greater mobilization of POPs from primary and secondary sources as well as increased airborne transport. It is unclear to what extent higher temperatures will accelerate the degradation of POPs, but the melting of ice in which they have been held for decades contributes to rising amounts of POPs and other pollutants in the environment.

Pesticides Including POPs

Pesticides are compounds designed to kill specifc pests but often reach non-target organisms as well. In one study, more than 90 per cent of sampled water and fish were found to be contaminated by several pesticides and estimates indicated that about 3 per cent of exposed agricultural workers sufer from an episode of acute pesticide poisoning every year. It is therefore imperative to know the nature of exposure and causes of contamination, and to identify action that can be taken to reduce pesticide levels in terrestrial and aquatic ecosystems. Long-term pesticide sales data constitute the main global and regional indicators of pesticide use. The last 25 years have seen a reduction in insecticide sales due to mammalian toxicity concerns, although general pesticide sales increased from US$5.4 billion in 2004 to US$7.5 billion by 2009 in the Latin American region, with 2,4-D, paraquat, methamidophos, methomyl, endosulfan and chlorpyrifos accounting for a high proportion of these sales.

Globally, the main 15 pesticides found in streams and groundwater include the herbicides atrazine and di-ethylatrazine, metolachlor, cyanazine and alachlor, and the insecticide diazinon. However, regarding fish, riverbed sediments and soils, the main pesticides still include persistent insecticides, heavily used in the 1960s and currently banned in most developed countries, such as DDT, dieldrin and chlordane. Moreover, endosulfan sulphate, the metabolite of endosulfan still in use in many countries, is a very common contaminant of surface and groundwater. Although the use of most organochlorine insecticides came to an end 10–25 years ago, they remain in the environment at levels of concern.

More than 70 per cent of the populations of low-income countries live in rural areas, and 97 per cent of rural populations are engaged in agriculture. While developing countries account for just one-third of global pesticide use, the vast majority of pesticide poisonings occur in these countries.

The extent of human exposure and the health effes of pesticides under future climate change conditions will depend on the adoption of less toxic practices that take account of changes in factors such as temperature and precipitation.

Obsolete Pesticides

Pesticides become obsolete when they can no longer be used for their intended purpose. There are four major international agreements for their regulation: the Stockholm, Rotterdam and Basel Conventions and the 1998 Protocol on POPs to the 1979 Geneva Convention on Long-Range Transboundary Air Pollution (UNECE Geneva Convention 1979/98). It is difficult to estimate exact quantities of obsolete pesticides because many are very old and documentation is scarce. However, amounts of obsolete pesticides that do not fall under the Stockholm Convention remain vague and can only be roughly calculated. On the basis of experience in Africa and the Middle East, UNEP estimates that on average POP pesticides make up only around 30 per cent of all existing obsolete pesticides.

Country-by-country assessments carried out by the International HCH and Pesticide Association suggest that obsolete pesticides could amount to between 256 000 and 263 000 tonnes in the countries of the former Soviet Union, the Southern Balkans and new Member States of the European Union (defned as EU-12, EU accession countries, the countries of the European Neighbourhood Policy (ENP), the Russian Federation and Central Asia together), costing approximately US$780 million to dispose of, while some estimates for Africa by UNEP Chemicals suggested that there may be as much as 120 000 tonnes remaining, costing some US$200–250 million to dispose of, applying UN Food and Agriculture Organization cost estimates. These assessments alone identify 376 000–383 000 tonnes for disposal at a cost of US$968–1 040 million.

The Africa Stockpiles Programme (ASP), launched in 2005, aimed to clear all obsolete pesticides and contaminated waste in Africa within 10–15 years and to promote prevention measures and capacity building. It is very likely that the costs of inaction by far exceed the costs of cleaning up. As underlined by the European Environment Agency (EEA), downplaying the costs of inaction is a frequent phenomenon and analysis suggests that the costs of inaction are high.

Metals, Metalloids and Heavy Metals

Inorganic pollutants, including metals and metalloids, also adversely impact human populations on a global scale. Unlike organic chemicals, metallic elements do not degrade and may accumulate in the environment and become increasingly bio-available over time. Their impacts are often most severe in the developing countries where they are mined, processed, used and recycled with limited environmental control and regulation. Populations in more developed countries also sufer from historic and on-going industrial emissions of pollutants, as well as from associated releases of other pollutants such as sulphur oxides, which cause acid rain, and acid mine drainage. Contamination even extends to Antarctica, as industrial pollutant emissions are carried there by long-range atmospheric transport from other continents. Pollutants can also be re-released after decades as glaciers melt.

Poisoning by naturally occurring arsenic is a global problem. More than a decade ago it was estimated that 130 million people around the world have been exposed to toxic levels of arsenic in drinking water, above the WHO recommended limit of 10 parts per billion, but there is mounting evidence that arsenic toxicity occurs at levels below that standard. There are also many unexplored sources of arsenic and the total number of people affeed may be higher. Associated toxicities include diabetes and skin, kidney, lung, neurological and vascular diseases – most notably blackfoot disease which leads to gangrene – and bladder cancer. These diseases are most prevalent in vulnerable populations living on subsistence diets of arsenic-contaminated foods and with limited access to clean water, minerals and nutrients, which partially counteract that toxicity. Arsenic pollution in Bangladesh, which resulted from drilling wells to protect the population from surface waters contaminated with pathogens, has been described as "the largest poisoning of a population in history". Populations in both developed and developing countries may be exposed to arsenic at contaminated sites left behind by the formerly widespread use of arsenic as a pesticide.

Lead is among the most prominent of the global contaminants, with several activities being responsible for acute lead poisoning. There are on-going human health problems at previous mining and smelting sites, including in Kabwe, Nigeria and the Rudnaya River Valley, Russia, where high lead levels in children persisted after the smelters in both areas were closed, and in La Oroya, Peru, where 99.7 per cent of the children living nearest the smelter were found to have dangerously high levels of lead in their systems.

On a global scale, some 85 per cent of lead-acid batteries are recycled, but there are recycling sites such as in Dakar, Senegal, where the average blood-lead concentration of children was at 130 micrograms per decilitre, enough to cause acute toxicity or even death. Children may also be exposed to lead in paints, which has been phased out in developed countries but persists in some developing ones. Electronic waste recycling can also involve exposure to lead in solder, and there are sites such as Guiyu, China, where 82 per cent of the children tested in the village had blood-lead concentrations above the US Centers for Disease Control action level of 10 micrograms per decilitre. Although that level is two orders above the estimated natural lead level, there is no established lower threshold for lead toxicity in humans.

Most coals contain tiny proportions of mercury, so industrial mercury fluxes to the biosphere are projected to increase with greater fossil fuel combustion. While large amounts of mercury are thus released into the environment from numerous industrial activities, reports of acute neurotoxicity from mercury poisoning are now primarily associated with its use to amalgamate gold in artisanal mining, which is practised in more than 50 countries. In Indonesia and Zimbabwe, all of the children tested in two mining areas were found to have elevated mercury levels and corresponding signs of mercury intoxication, whether they were directly involved in mining or not. This poisoning of children is of special concern because mercury, even at sub-lethal levels, is a neurotoxin that can permanently impair development and – like some other toxins – foster auto-immune resistance, making children and adults more vulnerable to infection and disease, as has been found with artisanal gold miners in Brazil. Currently, UNEP is convening an intergovernmental negotiating committee to prepare a global legally binding instrument on mercury: more than 100 countries are participating and a global treaty text is expected to be ready for adoption in late 2013.

A number of other metals such as zinc, copper and manganese could have harmful human and environmental impacts at certain levels. Cadmium, which was once used in pigments and for electroplating, is the most toxic, and contaminated sites may remain. Its main uses now are in rechargeable nickel-cadmium batteries, and collection and recycling of these items must be efcient if it is not to be released to the environment. Cadmium is also released into the environment by some fossil fuel combustion, and is in addition a natural contaminant in phosphate deposits, so may be transferred in fertilizers and taken up by root vegetables.

Radioactive Materials

Radioactive material has been in use since the 1890s, increasing significantly with the advent of nuclear energy in the 1940s and its exploitation in weapons, with a concomitant increase in the generation of radioactive waste and contaminated sites. In addition, the use of radioactive materials in industry, research and medicine continues and increases, as do the mining and processing of minerals containing elevated concentrations of naturally occurring radionuclides. Some contaminated sites have been remediated at significant cost, while others remain to be addressed. The rising cost and reduced availability of fossil fuels have from time to time favoured the adoption of nuclear power, as have recent concerns over greenhouse gas emissions. However, social attitudes to nuclear accidents such as those at Three Mile Island and Chernobyl – which are rare but can have a very high impact – have exerted a restraining influence. It was predicted in 2008 that the use of nuclear energy would increase by 15–45 per cent by 2020 and by 25–95 per cent by 2030, but future activity is likely to be affeed by responses to the more recent Fukushima disaster.

Radioactive waste takes many physical and chemical forms and has difering radioactive properties. The international system of classification links waste classes (exempt, very short lived, very low level, low level, intermediate level, high level) to options for management and disposal. Disposal is the fnal step in the management of radioactive waste, generally in near-surface or deep land-based facilities. Apart from high-level and some intermediate-level waste, the majority has been disposed of in such facilities.

About a hundred near-surface facilities exist, and others for disposal of waste of various levels are under development in a number of countries, although the process of selecting and designing a site is often contentious. Many nuclear reactors are ageing and will need to be decommissioned in the near future, resulting in radioactive waste and signalling the need for disposal facilities and trained professionals to operate them. As of 2 February 2012, 435 nuclear power reactors with a combined capacity of about 368 gigawatts are in operation in 30 countries, of which around 75 per cent are more than 20 years old, and 63 plants with a combined capacity of 61 gigawatts are under construction in 14 countries.

Contracting Parties to the Joint Convention on Radioactive Waste and Spent Fuel increased steadily after its establishment in 1997 to number 58 in April 2011, and are committed to ensuring a high level of safety in radioactive waste management. At the 2009 triennial review meeting, the reports of 45 Contracting Parties were reviewed with the conclusion that there is a commitment to improve safety, make progress in building, maintaining and implementing legal/regulatory frameworks, and observe good practices in national radioactive waste management strategies and policies. Despite progress since the 2006 review meeting, however, the 2009 meeting concluded that much still needed to be done to meet the following challenges:

- implementation of national policies for the long-term management of spent fuel, including disposal;
- siting, construction and operation of spent fuel and radioactive waste disposal facilities;
- management of legacy wastes;
- monitoring of disused sealed sources and recovery of orphan sources;
- knowledge management and human resources development; and
- provision of financial resources for liabilities.

There has been a growing trend for disposal facilities to undergo international peer-reviewed safety demonstration. In addition, the 2010 General Conference of the International Atomic Energy Agency (IAEA) created an International Working Forum on Regulatory Supervision of Legacy Sites, aimed at enhancing regulatory regimes, the professional development of regulators and the application of safety and environmental assessment.

Emerging Issues

Policy making and regulatory processes are naturally prone to lag behind rapid changes taking place in the global production and distribution of chemicals and wastes. The challenge is to protect human health and the environment from the undesirable effes of chemicals and wastes even when there are inadequate quantitative data and the potential life-cycle hazards of both old and new materials are incompletely understood.

Nanomaterial and Nanoparticles

Many new materials are produced as minute particles of a nanometre - or one-billionth of a metre - in size, and they exhibit chemical and biological properties that are quite different from those of the corresponding bulk materials. Commercial applications of nanomaterials include, for example, food packaging, personal care products, cosmetics and pharmaceuticals. Their unique properties make nanomaterials useful in cancer therapies, the neutralization of pollution or improvement of energy efciency. However, safety testing is in its infancy and governments have been slow to adapt existing regulations to these new materials, even though they are widely marketed and some potential for human exposure has been identified. More research is needed for a better understanding of workplace and consumer exposure and related impacts on human health, especially as some of these materials are known to pass through the skin and are small enough to penetrate cell membranes and cause toxic effes at cellular and sub-cellular scales. Furthermore, very little is known about whether nanomaterials or nanoparticles are released from products when they are incinerated, buried or degrade over time, so it is possible that they will pose a serious waste disposal challenge. Sound decision making on nanotechnology has provoked much debate among developed-country regulators, and is increasingly doing so among the regulators of developing countries.

Plastics in the Environment

Plastics are ubiquitous in the environment. They are widely used in many products and have many formulations. The simple plastic bag is a prime example of how a utilitarian object can become an environmental hazard. More than 500 billion plastic bags are used every year but many are improperly disposed of, ending up as marine litter. This significant problem was highlighted in the *UNEP Yearbook 2011,* showing that discarded plastic debris forms a major component of marine litter, degrading into micro-pollutants in ocean gyres, fouling beaches in coastal waters, and entering the food chain where it is consumed by marine fauna such as turtles and sea birds, weakening or killing them by affeing their digestion, respiration and reproduction. There is concern that these plastics also act as transport vectors of persistent organic pollutants such as PCBs and similar compounds, with chronic effes on wildlife. The solution is sound management, preventing the escape or discharge of this material, yet rates of plastic recycling and reuse vary greatly, from more than 80 per cent in some EU countries to only a small percentage in many developing ones. The Global Programme of Action (GPA) for the protection of the Marine Environment from Land-based Activities and other local and regional initiatives are seeking to address this issue.

Electronic Waste

The high turnover of equipment in the information and communication technology industry has caused an increase in obsolete electrical and electronic products, which in turn has generated almost uncontrollable volumes of end-of-life products driving a global trade in e-waste. As the fastest growing waste stream in the world, estimated at 20-50 million tonnes per year, e-waste has become one of the major environmental challenges of the 21st century. Generated by a wide range of electrical products, it is of particular interest because it contains not only hazardous substances including heavy metals such as mercury and lead, and endocrine-disrupting substances such as brominated fame retardants (BFRs), but also many strategic metals such as gold, palladium and rare earth metals that can be recovered and recycled. E-waste can thereby serve as a valuable source of secondary raw materials, reducing pressure on scarce natural resources and the environmental footprint of the mining industry.

Developing countries nonetheless remain the destination of most of the e-waste exported from developed countries as second-hand or used equipment. Yet these countries often lack the infrastructure, capacity and resources for its sound management, with the informal sector and vulnerable groups employing crude processing methods such as open-air burning or acid leaching to recover valuable metals like copper and gold. In the process, toxic substances in the waste may be released into the environment, posing a high risk to ecological and human health. Recent studies have revealed that by 2016, developing countries will generate twice as much e-waste as developed countries, but while electronic equipment has positive impacts on development and progress, it can have negative impacts on both human health and environmental integrity as end-of-life e-waste. This is a growing environmental and public health issue that threatens attainment of the Millennium Development Goals (MDGs) in developing countries and economies in transition.

Endocrine Disruptors

Endocrine disruption is the term given to the alteration of hormonal signals in living systems when they are exposed to chemical substances. A considerable number of chemicals have been shown to be endocrine disruptors, affeing the growth and reproductive and neurological development of many species, including humans. In addition, the numerous chemical substances, both natural and anthropogenic, that are present in the environment in low concentrations come together to exacerbate exposure of both humans and wildlife. Many investigations have been conducted since publication of the *Global Assessment of the State-of-the-Science of Endocrine Disruptors,* and it is clear that both inorganic and organic substances can affect hormonal signalling. UNEP has proposed listing this as an emerging policy issue for listing under the Strategic Approach to International Chemicals Management (SAICM).

Open Burning

In open burning, the pollutants produced by combustion are released directly to the air and so enter the environment in uncontrolled ways. Open burning can include forest wildfres, planned combustion activities such as burning stubble in preparation for a subsequent grain crop, irresponsible burning of waste such as domestic rubbish and e-waste, arson-initiated combustion of scrap tyres, and even public detonation of freworks. Polycyclic aromatic hydrocarbons (PAHs) are always released in these processes, and (in the case of freworks) heavy metals such as lead and copper are also released. PAHs are widespread in the environment in both developed and developing countries, and concern about their carcinogenic properties has led to their classification as primary pollutants by agencies such as the US Environmental Protection Agency.

Gaps in the Understanding of Chemical Toxicity

Since humans are continually exposed to a multitude of manufactured chemicals, there is a need to understand the behaviour of these chemicals and their interaction with human health and the environment. Previously unsuspected properties of widely used chemicals present legacy problems that raise concern in the scientifc community and among the public. For example, of the many chemicals that have been found to have endocrine-disrupting properties, bisphenol-A is present in many plastic baby bottles and food can liners, and phthalate esters in various flexible plastics including some children's toys. Consumer vigilance is not enough to prevent exposure in such cases because the presence of these chemicals is usually not evident to the non-expert. This places a heavy responsibility on public authorities to inform people about potential risks associated with manufactured chemicals, and on manufacturers to exercise the extended or individual producer responsibility approach and to search for alternatives.

Most existing chemical regulations worldwide address the effes of individual substances. Managing single chemicals is difficult enough, but there is also concern about gaps in understanding human exposure to mixtures of chemicals. As mentioned, little has been done to study the toxicology of mixtures. There is an urgent need for further risk assessment of the combined exposure of multiple chemicals – the chemical cocktail or synergistic effes – to human health and the environment. Integrated environmental risk assessments based on state-of-the-art dynamic pollutant modelling and toxicological experiments on chemical cocktails will help quantify planetary boundaries for chemical pollution.

Chemical Properties, Patterns of Use and The Environment

There is a lack of information about the health and environmental effes of many chemical substances and about the products in which different types of chemicals are used. Huge gaps in the assessment of chemicals arise from two causes. First, many were introduced and became established items of commerce before systematic assessments began. Where there is mounting evidence of harm or potential harm, action can lead to regional controls and eventual listing under global conventions, but most industrial chemicals remain unassessed. Second, concerns have arisen over hitherto unsuspected properties such as the endocrine activity of phthalates and bisphenol A, for example, or long-range transport coupled with bioaccumulation. Furthermore, academic assessment suggests the potential for further industrial chemicals and pesticides to qualify as POPs. It should also be noted that wastes are often mixed, which makes it extremely difficult to assess the risks of any chemicals present. In addition, residues arising from the recycling of hazardous waste may contain a higher concentration of toxic materials than the recyclable materials themselves.

Long-term monitoring programmes for POPs in the environment as well as in human tissue need to be maintained and expanded, in particular in the southern hemisphere. They are essential for a better understanding of trends in global chemical pollution and for the Stockholm Convention's evaluation of effectiveness.

More extensive work on chemical toxicity inventories is aiming to fll a significant gap. An example is the European legislation on Registration, Evaluation, Authorisation and Restriction of Chemicals (REACH). This has extended the number of chemicals covered by regulations, notably those that were on the market before 1981 and previously exempt.

Limited information on chemicals in products makes it difficult to document the extent of the risk posed to human health and the environment. Initiatives such as the ongoing UNEP Global Chemicals Outlook and the Cost of Inaction (UNEP Mainstreaming of Chemicals) will help to fll some important knowledge gaps.

In addition to scientifc knowledge gaps, sound chemicals and waste management is also hampered by a lack of resources, capacity and compliance monitoring. A lack of education and training also limits appropriate management of chemicals and wastes in many developing countries. Increased trade resulting from free trade agreements can complicate this picture, as such agreements may well exert even more pressure on emerging economies with respect to regulating or restricting chemical use.

Chemicals, Wastes and Drinking Water

At the global level, about 1.1 billion people do not have access to a safe water supply and 2.6 billion people do not have access to adequate sanitation facilities. The associated health impacts are alarming: 1.7 million deaths per year, of which 90 per cent are children under five years of age. The costs of water pollution may represent between 0.3 and 1.9 per cent of rural gross domestic product (GDP). Industrial sectors with the potential for significant water pollution include the chemicals sector, food and beverage sector, textile and mining industry and pulp and paper sector. The policy framework for regulating the industrial point sources of water pollution is well developed in most OECD countries, although some pollutants such as heavy metals and chlorinated solvents remain a concern. Increasing attention is being paid to non-point sources, such as agricultural run-of, which are more difficult to regulate but can lead to nitrate pollution of water bodies. In addition to efforts to reduce the run-of of organic pollutants from fertilizers and manure, organophosphates from pesticides are also a concern. Studies reviewed by the OECD suggest that national measures to reduce agricultural run-of and manage storm water, including targeted measures to reduce a variety of different pollutants such as arsenic and nitrates, could yield health benefits in excess of US$100 million in large OECD economies. In non-OECD countries, the cost of inaction with respect to unsafe water supply and sanitation is particularly acute.

Reinforcing a Global Response

The Basel, Rotterdam and Stockholm Conventions and other instruments that address chemicals and wastes – including the Montreal Protocol on Ozone-Depleting Substances, MARPOL, the London Convention, and regional treaties like the Bamako, Waigani or Mediterranean Conventions, as well as the future Minamata Convention on Mercury – represent the foundation on which to build and consolidate a global response to protect human health and the environment from the adverse effes of chemicals and waste. Discussions conducted under the auspices of these global instruments enable emerging problems to be foreseen and facilitate the formulation of ways to manage issues soundly and collectively on a sustainable basis. All these global legally binding instruments, as well as regional agreements such as those agreed by the OECD and the European Commission, share the universal principle of an environmentally sound management of chemicals and waste. A key feature of this global architecture is transparency in the collection and dissemination of information. The EU chemicals legislation, REACH, is exemplary of such efforts. But large gaps remain, both in addressing the number of chemicals and nanomaterials present in the market, and in the fact that many countries are unable to manage hazardous chemicals and waste in an environmentally sound way.

With the Basel, Rotterdam and Stockholm Conventions sharing the common objective of protecting human health and the environment from hazardous chemicals and wastes, the Parties to these agreements have embarked on rationalizing their operations to improve assistance to countries in managing chemicals at different stages of their life cycle. This has been exemplified by the establishment of the International Panel on Chemical Pollution (IPCP) in 2008, enhanced cooperation and coordination between the three conventions during their respective Conferences of the Parties in 2008 and 2009, and their simultaneous extraordinary meetings in Bali, Indonesia, in February 2010. Since early 2011, the convention secretariats have been working under a joint Executive Secretary, opening up the possibility of a more holistic approach to the sound management of chemicals and waste.

References

Basel Convention (1989). *The Basel Convention on the Control of Transboundary Movement of Hazardous Wastes and their Disposal.* http://www.basel.int/

IAEA (2009a). *Classification of Radioactive Waste General Safety Guide.* Series No. GSG-1. International Atomic Energy Agency, Vienna

Stockholm Convention (2001). *Stockholm Convention on Persistent Organic Pollutants.* Adopted 2001. Secretariat of the Stockholm Convention, Chatelaine.

UNCED (1992a). *Rio Declaration on Environment and Development.* United Nations Convention on Environment and Development, Rio de Janeiro

UN-Habitat (2010). *Solid Waste Management in the World's Cities: Water and Sanitation in the World's Cities 2010.* United Nations Human Settlements Programme and Earthscan, London and Washington, DC

7

Governance of Earth System

The first pictures of the Earth from space stimulated an immediate and profound appreciation of its fnite boundaries. Scientifc advancements continue to enable a better view of the Earth as a whole. This includes a combination of surface and remote-sensing global observation systems that can document global-scale phenomena, advances in the ability to reconstruct past states of the environment, and enhanced computing power to conduct global-scale simulation experiments. Evidence shows that human activities are now so pervasive and profound in their consequences that they affect the Earth at a planetary scale.

The Earth System

A system is a collection of component parts that interact with one another within a defned boundary. The Earth System is a complex social-environmental system, including the vast collection of interacting physical, chemical, biological and social components and processes that determine the state and evolution of the planet and life on it. The biophysical components of the Earth System are often referred to as spheres: atmosphere, biosphere, hydrosphere and geosphere. They provide environmental processes that regulate the functioning of the Earth, such as the climate system, the ecological services generated by the living biosphere, including food production, and natural resources like fossil fuels and minerals. Humans are an integral part of the Earth System. All spheres include countless subsystems and levels of organization. The interactions within and between these spheres are complex and the predictability of future states of the Earth System is limited.

Unprecedented Changes

Some experts suggest that the Earth has entered a new geological epoch, the Anthropocene. The word was coined by Nobel Laureate Paul Crutzen to capture the idea that humans are now overwhelming the forces of nature. An implication of entering the Anthropocene would be the leaving of the Holocene, the interglacial period that has provided humanity, over the past 10 000 years, with extraordinarily good living conditions, enabling the development of modern societies and a world with 7 billion people.

Crutzen suggests that the Industrial Revolution 250 years ago saw the beginning of the Anthropocene. The unprecedented rise in human population since the early 19th century, from less than a billion to 7 billion at present, is inherent to the Anthropocene as it unfolds. Many societal changes have accompanied this proliferation of the human population, such as increased consumption of natural resources and an enormous expansion of dependence on fossil fuels.

The Earth System demonstrates complexity in its natural variability independently of, and previous to, human influence.

Ice cores in Antarctica have shown that during the past 800 000 years, air temperature and carbon dioxide (CO_2) concentrations have oscillated within a relatively limited range, with variations that could be largely linked to factors such as the irregularities of the Earth's rotation and motion along its orbit around the Sun. Current concentrations of atmospheric CO_2 are, however, well outside the range of the past, having risen from 310 parts per million (ppm) in 1950 to 391 ppm in 2011, with half the total rise in atmospheric CO_2 since the pre-industrial era having occurred in the last 30 years.

Biodiversity, the variety of life on Earth, has evolved over the last 3.8 billion years or so of the planet's approximately 5-billion-year history. Five major extinction events have been recorded over this period, but, unlike the previous events – which were due to natural upheavals and planetary change – the current loss of biodiversity is mainly due to human activities and is often referred to as the sixth global extinction. According to the *Global Biodiversity Outlook 3,* the abundance of some vertebrate populations fell by nearly one-third on average between 1970 and 2006 and continues to fall globally. Many biologists consider that coming decades will see the loss of large numbers of species, increasing the risk of abrupt change in landscapes and seascapes. Fewer scientists appear to have recognized that, in the longer term, these extinctions will alter not only biological diversity but also the evolutionary processes by which diversity is generated.

Given the interconnections between the different spheres of the Earth System, changes in one part of the system have effes in one or more of the others.

Earth System Complexities

The complexity of the Earth System is associated with its countless interacting processes, at many scales and levels of system organization. Importantly, these interactions mean that changes rarely occur in linear and incremental ways. Instead, the dominant behaviour when the various systems on Earth undergo change is for it to happen in a non-linear way, driven by feedbacks that either dampen change (negative feedbacks) or reinforce it (positive feedbacks). Many such feedbacks shape the Earth System.

Positive feedbacks are increases in system reaction that may destabilize the system and move it into another state – a regime shift. An example of a positive feedback is the effect of black carbon deposition in the Arctic. Black carbon particles are emitted into the atmosphere from the incomplete combustion of biomass and fossil fuels. The Arctic climate is especially vulnerable to black carbon deposition because of its impact on the albedo (reffeivity) of snow, glaciers and sea ice. Black carbon makes the surface darker, thus reffeing less radiation, which leads to an increase in warming and ice/snow melt. Ramanathan and Carmichael report that, at high elevations in the Himalayas, positive feedback from the increased absorption of solar radiation by black carbon may be as important in the melting of snowpack and glaciers as are the rising temperatures that result from increased atmospheric CO_2.

An important relationship between the temperature and carbon content of the Earth's atmosphere manifests itself both on relatively short and on geological timescales, and is the result of many contributing feedbacks in the atmosphere and other components of the Earth System. For example, with the higher temperature and greater acidity of ocean waters, the ability of the ocean to act as a carbon sink weakens. This positive feedback is one that increases system reaction and is thus destabilizing.

Another destabilizing feedback, increasingly discussed in climate science, is related to the carbon reservoirs in the Arctic permafrost. If rising temperatures lead to permafrost thaw, this will release carbon and lead to further increases in temperature and consequently to even more permafrost thaw and more releases of carbon.

The role of biodiversity in such feedback processes is not currently well understood because of the complexity of interactions in physical, chemical and biological processes. It is, however, well known that a positive feedback, which could enhance climate change, will occur if the carbon stored below ground is released to the atmosphere by accelerated respiration induced by soil warming.

A negative feedback is when the initial response is suppressed, which tends to be stabilizing. For example, if increased water in the atmosphere leads to greater cloud cover, this raises the percentage of sunlight reffeed away from the Earth (albedo), which leads to a fall in the temperature of the atmosphere and a decrease in the rate of evaporation.

So far, the dominant aggregate response of the Earth System to human pressure has been to dampen its impacts. This is explained by the inherent resilience of the Earth System, where the biosphere interacts with the climate system, in particular, to bufer disturbances including some induced by humans. As a result, as a negative feedback response to CO_2 emissions from human activities, the global carbon sink in the biosphere has increased from approximately 2 billion tonnes of carbon per year in the 1960s to approximately 4 billion in 2005. However, there are indications that the capacity of the biosphere to bufer pressures from global environmental change is declining, and there is growing evidence of positive feedbacks occurring at the local level, for example eutrophication of lakes, through to the regional level, such as accelerated melting of Arctic ice cover due to a regional amplification of warming.

Earth System Changes and Implications for Human Well-being

The key Earth System changes discussed above have impacts on the environment, on economies and on societies. Illustrative and by no means comprehensive examples of these impacts follow, showing the interconnectedness of the Earth System and the effes that human activities and environmental change have at all scales.

Polar Regions

Many of the complex changes in global environmental conditions tend to amplify in the polar regions. For example, heat flux from the lower latitudes leads to accelerating melt of sea ice as well as a loss of mass in Arctic glaciers and the Greenland and Antarctic ice sheets, all of which contribute to global sea level rise. There are many ways in which the polar regions affect the lower latitudes and the whole globe.

The Arctic

This amplification of global warming has been confrmed by instrumental records and reconstruction of past climates, and has been demonstrated in climate model simulations.

The amplification is caused by several factors including heat transport to the Arctic; melt of sea ice enhanced by black carbon depositions on snow and a corresponding reduction in albedo; an increase in long-wave (infrared) energy emitted downward by the atmosphere; and an increase in heat-absorbing black carbon aerosols in the atmosphere. The rapid shrinking of Arctic sea-ice cover is part of the positive climate feedback. In addition to the reduction in the area covered by ice over the past 30 years, as shown by satellite data, substantial loss is also taking place in the oldest and thickest ice. The disappearance of sea ice, which acts as a thermal insulator between the ocean and atmosphere, results in an enhanced upward heat flux that warms the lower troposphere in the Arctic, affeing the general atmospheric circulation in a significant part of the northern hemisphere. This alters storm tracks, precipitation patterns and the conditions that lead to heat waves and cold spells. For example, the emerging atmospheric pattern of a warm Arctic Ocean and cold continents favours more frequent and severe Arctic air outbreaks during the cold season, affeing the well-being of hundreds of millions of people living in the mid-latitudes of the northern hemisphere.

The Arctic Council's *Snow, Water, Ice and Permafrost in the Arctic* assessment shows that temperatures in the permafrost have risen by as much as 2°C over the past two to three decades, particularly at colder sites. Warming in the Arctic causes permafrost thaw and loss, as seen in the increased depth of seasonally thawing soil above the permafrost in

Scandinavia, Arctic Russia west of the Urals, and inland Alaska; the 30–80 km northward retreat of the southern limit of permafrost in Russia between 1970 and 2005; and the 130 km retreat in Quebec, Canada, during the past 50 years. A regional process of permafrost thawing supports increased microbial activity and is likely to lead to the release of carbon that is currently locked up in frozen soils, initiating a global positive climate feedback. It is possible that by 2030 the Arctic will become a carbon source rather than a carbon sink.

The warming and opening of Arctic waters also has implications for the accessibility of hydrocarbons and other natural resources. The production of oil and gas and increased shipping could turn the Arctic into an area of rapid industrial development, leading to additional anthropogenic emissions of carbon. This is another example of a positive climate feedback, involving both natural influences (the greenhouse effe) and social ones (human activities).

Another unexpected manifestation of the global and regional links in the Earth System was observed in the Arctic in boreal spring 2011. An unprecedented stratospheric ozone loss of approximately 80 per cent at altitudes of 18–20 km was attributed by Manney *et al.* to anomalously long-lasting cold conditions in the Arctic's lower stratosphere, which in turn led to a persistent enhancement of the atmospheric content of ozone-destroying forms of chlorine.

The Antarctic and the Southern Ocean

This remote region is still poorly understood and there is limited capacity to observe the highly complex Earth System interactions that take place there. Numerous observations show that Southern Ocean waters are warming more rapidly than the global ocean average. Warming of intermediate waters was reported by Gille, while a comparison of ship and foat observations showed widespread warming and freshening of the Antarctic Circumpolar Current waters. Abyssal and deep-water measurements also indicate warming trends. A profound peculiarity of this region is the stratospheric ozone hole, which has had a significant impact on the Antarctic environment over the last 30 years, altering the main regional pattern of climate variability, the Southern Annular Mode and associated winds, which tend to shield large parts of the continent, except the Antarctic Peninsula, from greenhouse-gas induced warming.

The Antarctic is the Earth's largest frozen store of freshwater, with the potential to cause sea level rise equivalent to 61.1 metres. While significant parts of the Antarctic ice sheet rest on land, these areas are still below the current mean sea level. For example, the ice body of the West Antarctic ice sheet is in many places more than 1 000 metres below the ocean surface. Recent estimates indicate a potential contribution to global sea level of 3.3 metres by this ice. Recent airborne geophysical measurements over previously unexplored areas of the East Antarctic ice sheet have shown that it too is largely resting below sea level. There are therefore concerns about marine ice sheet stability in a rapidly warming climate. While recent regional temperature trends in Antarctica have not been very significant and in some locations negative, the Faraday/ Vernadsky Station in the northwestern part of the Antarctic Peninsula has observed an increase of 0.53oC per decade for the period 1951-2006. This local warming and the accompanying changes in winds are considered to be the main causes of the collapse of the Larsen ice shelves A in 1995 and B in 2002. The potential destruction or accelerated melt of the West Antarctic ice sheet under current warming is a subject of intensive research.

As a general rule, when atmospheric concentrations of CO_2 increase, the oceans tend to absorb more of it. However, in the Southern Ocean, which represents a significant fraction of the global ocean carbon sink, there is a declining ability to absorb CO_2. One of the likely reasons for this is a 15-20 per cent intensification of the circumpolar westerly winds over the Southern Ocean since the 1970s, which can be partially attributed to the effes of the stratospheric ozone hole. This phenomenon also has important implications for Antarctic biodiversity.The ensemble of climate-chemistry models of the World Climate Research Programme Chemistry-Climate Model Validation project (CCMVal2) indicates that complete recovery of the stratospheric ozone layer resulting from implementation of the Montreal Protocol should occur around the middle of this century. However, the ozone layer restoration may affect the Southern Annual Mode and associated winds, weakening the currently existing restraint on greenhouse gas-induced warming in Antarctica and the Southern Ocean, and potentially leading to other significant local-to-global changes.

Implications for Human Well-being

The patterns of change in the polar regions and their links with activities and impacts elsewhere on the globe illustrate the archetypical pattern of vulnerability described in *GEO-4* as "misusing the global commons". This misuse leads to the exposure of people and the environment to resource depletion, for example dwindling fisheries or land in the case of sea level rise, and to environmental transformations such as climate change and sea level rise. Those who are most vulnerable to changes resulting from misuse of the global commons are usually not responsible for the misuse itself.

From an Earth System perspective, the ongoing and potential future changes of the polar regions – given the sometimes long timescales, the interconnections with the rest of the globe, the interactions between problem areas like stratospheric ozone recovery and global warming, and possible catastrophic events such as the melting of the West Antarctic ice sheet – point to the need for holistic responses to managing the global commons in order to reduce the vulnerability of people and the environment to potentially very large pressures.

The Hindu Kush-Himalaya

The Hindu Kush-Himalaya, sometimes referred to as the Third Pole, is one of the most dynamic and complex mountain systems in the world. It contains the largest amount of snow and ice found outside the polar regions, including more than 100 000 km^2 of glacier cover, and the sources of ten of the largest rivers in Asia. This mountain system, stretching 3 500 km through some of the world's wettest and driest environments, rising 8 km vertically through nearly every life zone existing on Earth, and at the geographical centre of the largest and densest concentration of humans, is recognized as an extremely fragile environment and particularly vulnerable to global warming.

Extreme vulnerability to natural hazards in the countries of South Asia is cyclical and repeatedly causes major setbacks in the socio-economic and equitable development of the region. Climate change is expected to increase both the frequency and magnitude of extreme weather events that lead to disasters, and calls for speedy action.

Uncertainties about the rate and magnitude of climate change and its potential impacts prevail, but there is no question that climate change is one of the many pressures that are gradually and powerfully changing the ecological and socio-economic landscape in the Himalayan region. This is particularly true in relation to water and ecosystem services, with significant implications for mountain communities and livelihoods, as well as downstream users, especially women who, for instance, often need to trek longer distances for potable water and fuel. In mountain areas, however, the influence of the changing climate has to be understood in the frame of overall changes due to modernization – communication, transport, infrastructure, monetization and others – and migration, which alters traditional gender relations.

Implications for Human Well-being

Climate change impacts in the Hindu Kush-Himalaya region and downstream are particularly severe due to the large number of people depending on climate-sensitive livelihoods such as agriculture. Here, more than 20 per cent live below the poverty line, amounting to around 260 million people. The International Food Policy Research Institute concluded that the negative impact of climate change on world cereal production may vary from 0.6 per cent to 0.9 per cent per year, but that in South Asia the impact could be as high as 18.2–22.1 per cent by 2080. Recent studies conclude that the Himalayan region and its downstream areas, including the Indo-Gangetic plains, the grain basket of South Asia, are also particularly vulnerable to climate change.

Poor and marginalized groups such as mountain populations and the inhabitants of downstream food plains are particularly vulnerable to climate change. The rough topography of the.

Himalayas combined with the precariousness of many homesteads on low incomes, makes the region a particularly food-sensitive area, with mudslides and unstable ground serious threats to settlement areas. Moreover, the risks of death and destruction are increased by the fact that people, after foods, often rebuild on the same risk-prone areas.

Mountain livelihoods are much more susceptible to environmental and economic upheaval than are livelihoods in the plains, and poverty in the mountains is exacerbated by climate change. Women, in particular, are vulnerable to the impacts of climate change and environmental degradation.

The Amazon

The Amazon forest is an extremely important component of the Earth System. It is the repository of the greatest terrestrial diversity of organisms on Earth, it exchanges vast amounts of water and energy with the atmosphere and thus affes local and regional climates, and it is a major carbon sink and reservoir containing 90 billion tonnes of carbon. This is about a ffifth of the total carbon contained in the world's tropical forests.

The Amazon has recently experienced two once-in-a-century droughts in the space of five years – 2005 and 2010. Both events caused rapid, widespread tree mortality leading to large increases in carbon emissions in undisturbed regions that are normally net carbon sinks. Even under declining deforestation rates, droughts increase vulnerability to fres by enhancing forest fammability and fire spread. According to model simulations by Vergara and Scholz, in a zero-deforestation scenario the effect of climate change alone, which is predominantly driven by fossil-fuel emissions, could reduce the extent of the Amazon biome by one-third by 2100. But when the effes of deforestation and fire are combined with those of climate change, the models show a much greater reduction.

Deforestation in general, and in the Amazon in particular, is a product of the interactions of multiple socio-economic factors in addition to the natural factors discussed above. The largest of these by far is conversion to cattle pasture to satisfy growing international demand for beef.

Implications for Human Well-being

In addition to its importance in regulating the global climate through acting as a carbon sink, the Amazon also provides livelihoods for both indigenous peoples and recent settlers. Foley *et al.* note that the Amazon system regulates freshwater and river fows, modulates regional climate patterns and controls the spread of vector-borne and water-borne diseases, all of which are crucial to human well-being. Farmers in the Amazon are vulnerable to climate change through the effes of drought, food and fire on planting times, the spread of disease, and impacts on food, water and human security.

Drylands

Desertification or land degradation in drylands is one of the greatest environmental challenges facing human society. The world's drylands – arid, semi-arid and dry sub-humid climatic regions, ranging from deserts to steppes and savannahs – cover approximately 40 per cent of the world's land surface area and are home to nearly 2 billion people. Degradation of a landscape is a particularly complex problem because it involves a tight coupling of socio-economic, meteorological and ecological processes.

Western Australia

One example of how changes in land cover can affect the regional climate in an arid zone is the rabbit-proof fence in Western Australia, built to prevent rabbits from damaging cropland and pastures. The fence spans more than 750 km and separates native vegetation to the east from 13 million hectares of cropland to the west. It did little to protect the crops from rabbits, but it illustrates how vegetation affes climate: there are more clouds and it rains more frequently on the east side of the fence where native vegetation remains. Several plausible explanations are provided by Nair *et al.,* who measured numerous physical and biological variables on both sides of the fence. They found that these variables difered substantially throughout the year on the agricultural side of the fence whereas only small seasonal variations were found over the native vegetation. Nair and colleagues concluded that the darker surface and greater roughness of the native vegetation resulted in an enhanced heat flux into the atmosphere, which increases the chances of cloud formation. Since measurements began in the 1970s, rainfall observations show about a 20 per cent decline in winter rainfall, confned mainly to agricultural areas.

The Sahel

The Sahel is a large semi-arid region that runs east-west across Africa south of the Sahara, and extends through ten countries. Rainfall is extremely variable and predominantly driven by two major, and undoubtedly interacting, factors: patterns in global sea surface temperature and large-scale changes in land cover that impact land-atmosphere interactions. The role of rainfall variability and vegetation dynamics in the Sahel has been the subject of many high-profle studies, and is particularly important because the population of Sahelian countries is projected to quadruple by 2020 relative to its population of 19 million in 1960.

Almost 6 000 years ago the Sahel was covered by grasslands and shrublands, with records of marine sediments and archaeological evidence showing a switch to arid conditions thereafter. More recently, there has been a marked shift from relatively wetter conditions with higher rainfall in the 1950s and 1960s to drier conditions in the 1970s and 1980s, followed by a general trend in increasing precipitation throughout the Sahelian region over the past 30 years, leading to what is generally referred to as a greening trend. Huber *et al.,* however, demonstrate the complexity of this trend, since vegetation changes are not always directly related to precipitation changes.

Implications for Human Well-being

In the case of Western Australia, land-use change brought unintended consequences. As well as the decrease in rainfall, the removal of deep-rooted native vegetation also led to a rise in the water table, increasing the surface salinity of the farmland and, hence, further decreasing agricultural productivity. As humans continue to clear land for agriculture a paradox is in the making: while food production may increase in the short run, it may be seriously decreased in the long run.

Another consequence of the widespread clearance of native vegetation in Australia for cattle raising and farming was its impact on indigenous peoples who had relied on previously abundant wildlife for their traditional diets. Many groups had little choice but to work on cattle stations and adapt to European foods. This has had a detrimental impact on their nutritional status and well-being, leading to chronic diseases associated with obesity.

Subsistence agriculture is the main source of household livelihoods in many parts of Africa, especially in dryland regions such as the Sahel. This constitutes a serious food security risk given the complex feedbacks between human activities, land cover and climate. The African Partnership Forum estimates that 75–250 million people living in the African drylands will be affeed by climate change.

Although a greening of the Sahel region is observed, rainfall in the Western part of the region has not increased. A study by Mertz *et al.* of 1 249 households in five Sahelian countries with annual rainfall ranging from 400 to 900 mm, found that climate factors, mainly inadequate rainfall, are believed by 30–50 per cent of households to be a cause of decreasing rain-fied crop production, whereas a wide range of other factors, such as changes of land tenure, was held responsible for the remaining 50–70 per cent. The diferences between the rain-fied crop and livestock sectors, as well as between the driest and wettest zones studied by Mertz *et al.,* illustrate the difficulty faced by people on the agricultural margins in the driest part of the Sahel who are trying to develop their rain-fied agriculture. Adaptation to climate change in the drylands will have to take these complex interactions into account.

Fires

The majority of global biomass burning occurs in the tropics where there are cycles of drought and exceptionally wet years. The African continent has the highest occurrence of vegetation fres, accounting for an estimated 30–50 per cent of the total annual biomass burned globally.

Large and uncontrolled fres have increased in the recent past on all vegetated continents and have caused tens of billions of US dollars of damage. Evidence from the United States and from Canada shows that fire extent in both countries increased significantly over the 20th century. In the western United States, the frequency of large fres has increased nearly fourfold and the extent more than sixfold since the mid-1980s. Recent fres in the Canadian and Alaskan tundra are unprecedented in the last 5 000 years. Satellite observations demonstrate a strongly non-linear relationship between climate and human activity: droughts result in more rapid deforestation while also reducing the fire bufering effect of peatland water tables, thereby increasing ecosystem vulnerability to fre. Projecting future fire dynamics is challenging because of the non-linearities in the various causal factors and the unresolved question of whether direct human activity or climate change plays a more dominant role in general.

Implications for Human Well-being

Among the significant effes of fres on human well-being are the destruction of assets such as homes; effes on human health and mortality, as demonstrated by the fres in Russia in 2010; and the lost livelihoods of rural resource-dependent communities, as experienced, for example, in Lebanon in 2007. In addition, the continuation or exacerbation of current global fire trends would have serious consequences for the enormous amounts of carbon stored in forests and other ecosystems, potentially unleashing an amplifying climate-carbon feedback and increasing the risk of dangerous climate change.

Shale Gas Basins

New proven technologies such as directional drilling and hydraulic fracturing have made the extraction of natural gas from low permeability geologic formations (shale formations) economically viable. These practices have accelerated the construction of new natural gas wells and their accompanying infrastructure – pipelines, roads, compressor stations and evaporation ponds – and led to extensive land fragmentation and disturbance, degraded air quality, and degraded surface-and groundwater quality. In parts of both the eastern and western United States, there has been rapid growth in the pace of development as new geologic targets become economical to drill using new technologies. Although to date such shale gas extraction has primarily taken place in the United States, it is expected to reach other parts of the world as use of the new technology expands, and as changing access, profitability compared to other gas resources, and shale gas characteristics make it viable.

While substituting the burning of coal with natural gas leads to emission reductions and may have some local air quality benefits, the impacts on air quality near concentrated natural gas development can be quite severe due to the release of hazardous air pollutants such as benzene, ozone-forming precursors and fugitive dust, among others. More broadly, such fuel switching, along with the continuing large-scale development and use of unconventional fossil fuels like shale gas, is likely to exacerbate human-induced climate change because the methane emissions are at least 30 per cent higher than from conventional gas. Furthermore, evaporation ponds used for the disposal of produced water – or coal seam gas water – in the western United States have recently been found to be major sources of volatile organic and hazardous air pollutants. The impacts of natural gas development on water resources are also broad-ranging, including the contamination of groundwater aquifers with potentially explosive levels of methane, of surface- and groundwater with chlorides, metals and organic compounds, and of streams where produced water is discharged, alongside high rates of consumptive water use for the drilling and completion of wells. The complex nature of many of the geologic formations from which natural gas is recovered may lead to many unknown impacts on groundwater resources. This is of particular concern because of the huge volume of shale gas deposits worldwide.

The chemicals used in hydraulic fracturing, as well as related surface water contamination and air pollution, are thought to be harmful to human health.

Overshoot

Scientifc understanding of the functioning of the Earth System and recent changes in it indicate a risk of crossing thresholds or tipping points, which would lead to fundamental state changes with major implications for human societies. Such shifts might include the transformation of rainforest to savannah or of hard to soft coral reefs, and changes in rainfall patterns. The risk of abrupt changes causing regional to global impacts – captured in such concepts as tipping elements and planetary boundaries – is a relatively recent insight from Earth System science. These frameworks for global sustainability complement previous concepts such as limits to growth, carrying capacity, ecological footprint and overshoot, which have in common an estimate of stocks of natural resources and critical loads of different pollutants predominantly in relation to health. Underlying all of these descriptors are different methods and assumptions for identifying points at which the ability of the Earth System to absorb anthropogenic changes is exceeded. While these methods and assumptions are still being discussed in the scientifc literature, their conclusions all point in the same direction: thresholds in the Earth System are being reached and the consequences are significant.

Forty years ago, Meadows *et al.* argued in *The Limits to Growth* that unchecked consumption and economic growth on a fnite planet were leading the Earth towards overshoot of its carrying capacity, which would be followed by major impacts on the global economy. Hall and Day looked back at the conclusions of this study and found that its warnings were generally correct. Turner compared historical data for 1970–2000 with scenarios presented in *The Limits to Growth* and found that 30 years of historical data compared favourably with key features of the business-as-usual scenario, which results in collapse of the global system midway through the 21st century.

The ecological footprint is used to understand human demand on the biosphere and the Earth's biocapacity. While there is still a need for improved data, humanity's ecological footprint overall has doubled since 1966, with major regional disparities. Of significance with regard to overshoot is the ecological footprint of megacities.

Another approach that points to Earth System limits looks at resource use. Material fow accounting, which quantifes all materials used in economic activities, accounts for the total material mobilized during extraction and for the materials actually used in economic processes measured in terms of their mass (tonnes). At the beginning of the 21st century, estimates of the quantity of global raw materials extracted ranged between 47 and 59 billion tonnes per year, with global annual material extraction having increased by a factor of eight during the 20th century.

Recognizing interactions and non-linear dynamics within the Earth System, the concept of planetary boundaries was introduced by Rockström *et al.* (2009a) to identify those key environmental processes that provide humanity with a safe operating space for well-being. Rockström *et al.*

identified nine planetary processes and proposed safe boundaries for seven of them – climate change, rate of biodiversity loss, the nitrogen and phosphorous cycle, stratospheric ozone depletion, ocean acidification, global freshwater use and change in land use. The proposed boundary position was placed at what was considered to be a safe distance from the risk of critical feedbacks and non-linear shifts that could trigger deleterious changes in critical environmental systems. The safe position for a process was based on an assessment of the current state of science, acknowledging that there is and will always be an uncertainty range for environmental risks. The safe boundary level for each environmental process was selected for the lower end of this scientifc uncertainty range to reffe a precautionary approach. The boundaries relate to rates and processes driven by human activities and not to resource scarcities.

The planetary boundaries framework indicates that environmental challenges at the global scale go well beyond climate change. Furthermore, evidence indicates that the processes analysed interact: transgressing one safe boundary may affect the distance from others. For example, expanding agricultural land may undermine the climate change boundary by increasing carbon emissions from terrestrial ecosystems. While the specifc numbers used in the planetary boundaries analysis may be challenged, the approach provides input to the debate about overshoot: early analyses indicate that humanity has already transgressed three boundaries – climate change, the rate of biodiversity loss, and global interference with the nitrogen cycle. A recent follow-up on human interference with the phosphorous cycle indicates that the phosphorous boundary has also been exceeded in freshwater systems.

Implications for Human Well-being

Ecosystems are essential for human well-being through their provisioning, regulating, supporting and cultural services. Human well-being refers to the extent to which individuals have the ability to live the kind of lives they value and the opportunities to achieve their potential, and is determined by a range of factors including access to resources, not just financial ones, security, good health and social relations (Introduction). All of these factors are affeed by changes in the Earth System. Global interconnectedness in the human-environment system also means that well-being in one place may be affeed by practices elsewhere.

From an Earth System perspective, it is important to consider the consequences for human well-being of exceeding the planet's carrying capacity or entering periods of abrupt and irreversible change. As shown in the examples below, the impacts of complex, non-linear changes in the Earth System are already having serious consequences for human well-being.

Multiple and Interacting Drivers affect Human Security

Climate variability and extreme weather influence food security. These drivers are complex and involve different pathways (regional water scarcity, salinization of agricultural lands, destruction of crops through food events, disruption of food logistics through disasters, and increased burden of infectious plant diseases or pests).

Crossing Thresholds: Significant Health Impacts

Land-use change and deforestation alter habitats by raising local temperatures and removing shade, changes that can facilitate the rapid development of malaria vectors. Pascual *et al.* demonstrated the importance of the well-recognized non-linear and threshold responses of malaria (a biological system) to the effect of regional temperature change.

Unprecedented Events Affect Assets and Human Security

Climate change already undermines human security and will do so increasingly in the future, by reducing access to, and the quality of, natural resources that are important to sustain livelihoods. In Bangladesh, for example, a significant number of people are affeed every year by riverbank erosion and foods that lead to loss of agricultural land, infrastructure and communication systems. These assets are essential for maintaining livelihoods.

Rapid Change and Indigenous Communities

As discussed, the Arctic is warming faster than elsewhere on the planet. Since 1975, temperatures in Alaska have increased by an average of 2.0–3.5°C. Approximately 200 indigenous villages along the navigable waters of Alaska's coasts and rivers are threatened by accelerated rates of erosion or fooding, and five communities have concluded that relocation is the only solution. Studies show that displacement has considerable cultural, social, economic and psychological impacts.

Transitions and Systemic Responses to Earth System Challenges

Earth System challenges have been characterized as "persistent problems of unsustainability" that are "... complex, ill-structured, involve many stakeholders, are surrounded by structural uncertainties, and are hard to manage". Persistent problems tend to reappear when only their symptoms are treated or when the measures taken are only marginal and incremental, and thus inadequate to deal with root causes.

The persistence of the problems is due to what Rotmans refers to as system failures:

- institutional system failures - a predominance of institutions that block innovation;
- economic system failures - inadequate market development or lack of investment capital;
- social system failures - entrenched behaviour;
- ecological system failures - regime shifts described earlier.

Transition Management

Addressing these system failures requires new, innovative forms of governance, including transition management. Ultimately, ignoring these systemic failures will result in non-linear, systemic and fundamental changes in the composition and functioning of the societal system, with changes in structures, cultures and practices.

Historical transitions, such as the post-1950 emergence of personal mobility, intensive agriculture or fossil energy infrastructure, were partly driven by the promise of solving societal problems such as poverty, inequality, lack of education and so on. However, these transitions have, in turn, produced their own problems. While individuals might now have access to cheap energy and mobility, the results are pollution, resource exploitation and congestion. The challenge in dealing with complex and persistent modern problems is to find new ways of dealing with them in a more anticipatory and exploratory manner. It is necessary to improve understanding of the dynamics of complex processes of change and try to influence their pace and direction.

A more thorough understanding of the forces that drive societal transitions is essential in policy making for the Earth System. While current mainstream policy and research approaches predominantly seek to improve existing systems, leading to gradual improvement, transitions thinking requires a fundamental shift. The current trend of marginal improvements and optimization of existing systems leads to lock-ins not only in technological systems but also in policy and consequently social systems that divert society from sustainability. Escaping such lock-ins requires radical shifts – transformative changes – that fundamentally alter structures, cultures and practices to achieve sustainability in the long term.

As transitions are increasingly likely to occur given the instabilities in both socio-economic and ecological systems, it is crucial that strategies be developed to influence these transitions effectively in terms of their speed and direction. While steering a transition in a command-and-control manner is not workable, it is possible to influence a transition by using various approaches, including coordinating existing social movements, niche-innovations and new practices in general.

Understanding the inevitability of transitions, and learning to govern and manage transition processes, is particularly important given the evidence of Earth System changes. New kinds of multi-level change processes are required that involve a dynamic interplay between gradually introduced, top-down changes and self-organizing bottom-up processes of social innovation, because traditional expert-driven, top-down approaches to problem solving are not flexible enough to address complex, non-linear and rapidly changing situations effectively.

These change processes require the active involvement of agents from science, policy, civil society and business, both in the development of new knowledge and in its application. The processes are necessarily iterative and involve developing a joint framing of a problem, a shared vision of the future, experimenting with solutions, evaluation and learning. Bottom-up solutions developed in this way should contribute to improving local sustainability and also reinforce the initial top-down changes and support their further extension. This is also proposed by the German Advisory Council on Global Change, which points to the need to empower the state to determine priorities and underline them with clear signals, while at the same time giving citizens more extensive opportunities to have a voice, to get involved in decision making and to take a more active role in politics.

With regard to bottom-up responses, Westley *et al.* point out that there are enormous reservoirs for learning and innovation that are often revealed in moments of crisis. Success involves listening to local communities for ideas, informing local populations of the resources and possibilities available, trusting them and allowing a diversity of innovative responses to emerge, as opposed to insisting on a top-down planning process.

Reflections

The Earth System is complex, with many interactions between and within subsystems, feedbacks and non-linearity. Humans, as an integral part of the Earth System, are changing it through their sheer numbers and their activities, although the impacts of these changes are not uniformly distributed, with some people and places more affeed than others. As a result of the enormous complexity of the system as a whole, it is not possible to predict the outcomes of rapidly increasing human pressures on the Earth System, but it is clear that thresholds have been or are being reached, beyond which abrupt and irreversible changes occur. These changes will affect the basic life-support functions of the planet.

References

Barnett, J. and Adger, W.N. (2007). Climate change, human security and violent conflict. *Political Geography* 26, 639655.

Eyring, V., Shepherd, T.G. and Waugh, D.W. (2010). *SPARC Report on Evaluation of Chemistry-Climate Models.* SPARC Report No. 5. Stratospheric Processes And Their Role In Climate. WCRP-132, WMO/TD-No. 1526.

Frantzeskaki, N. and Loorbach, D. (2010). Towards governing infrasystem transitions: reinforcing lock-in or facilitating change? *Technological Forecasting and Social Change* 77, 1292130.

Levin, S.A. (1998). Ecosystems and the biosphere as complex adaptive systems. *Ecosystems* 1,431436.

MA (2005). *Ecosystems and Human Well-being: Synthesis.* Millennium Ecosystem Assessment.Island Press, Washington, DC.

Sherman, K. and Hempel, G. (2008). *The UNEP Large Marine Ecosystem Report: A Perspective on Changing Conditions in LMEs of the World's Regional Seas.* United Nations Environment Programme, Nairobi.

8

Environmental Governance in Africa

Over the last generation, Africa has built a substantial record of responding to environmental challenges while addressing human well-being, and this provides a point of departure for strengthening policy and implementation.

As the policy appraisal shows, innovation by and partnerships between African people and their governments have underpinned this success, while the support of donors has been crucial to the implementation of some policies. The principles of the Paris Declaration on Aid Efectiveness – ownership, harmonization, alignment, management for results and mutual accountability – defne collaboration with donors and are designed to ensure that aid supports agreed government priorities and uses, and strengthens government systems rather than developing parallel institutions.

Despite tremendous progress, significant challenges remain, including population growth, rapid urbanization, climate change, unsustainable development choices and weak governance.

Africa's population – 1 billion in 2009 – is growing at close to 2.15 per cent per year, placing increasing demand on environmental resources. In 2010, some 395 million people, or 40 per cent of the total population, lived in urban areas. By 2040, the urban population is likely to be 1 billion and by 2050, 1.23 billion – 60 per cent of the total. In Africa's cities – characterized by extremes of prosperous centres and poor, informal settlements – many governments struggle to provide social services including access to water, to achieve food and energy security, and to manage environmental risks. Climate change and other adverse environmental change may increase urbanization and further strain governments' ability to cope, leading to greater instability.

Climate change – by exerting extreme pressure on ecological systems – is likely to increase the stress of vulnerable populations in urban and rural areas. More intense rainfall events contribute to more run-of and foods, threatening food security and settlements, while longer periods between rains and changing seasonal patterns contribute to crop loss. Sea level rise is likely to have significant impacts on coastal settlements, given the high population in potentially hazardous areas, including Akpakpa in Benin and Lagos in Nigeria, making climate-sensitive policies critical for conservation and adaptation. However, most existing policy lacks the framework to address the complex challenges of human vulnerability to climate change.

Policies and practices that originate outside the region can contribute to environmental change. External investments in land deals have increased rapidly as global demands for food and biofuels have risen – some 45 million hectares of such land investments, 70 per cent of the global total, are in Africa. Frequently, these investments have adverse impacts on land resources and livelihoods. Other examples of external drivers include climate change and waste disposal practices that have adverse impacts on land and water quality, and consequently on human health and food security. This makes the harmonization of policies between different countries as well as different regions a priority.

Over the last ten years, Africa's development approach has focused on securing growth through resource extraction, especially in the oil and mining sectors, and expansion of infrastructure. In the absence of strategic and integrated environmental assessment and robust accountability systems, this has led to environmental degradation. For example, weak accountability in decision-making systems has contributed to forest loss.

While governance and institutional arrangements vary, some challenges are of regional significance given the similarities across nations. Conficting laws, values and interests in and between countries adversely affect the ability to develop the collaborative institutional systems that are essential for managing ecosystems and responding to common challenges such as drought. At times, this disparity has resulted in resource conficts, including over the fair allocation of land and water. Inequitable sharing of the benefits and losses of natural resource management also reduces social-ecological resilience, often creating confict as deteriorating conditions in one sphere affect another. Weak land tenure, insufcient accountability and poor transparency compound this adverse reality. Sectoral planning that treats the environment as a set of separate resources rather than as a composite system further undermines environmental management. In this complex and unsatisfactory governance context, it is often vulnerable groups that sufer most.

Policy Appraisal

It appraises the selected policy options and shows that they are often mutually reinforcing, with positive impacts in more than one thematic domain. For example, developing more effective accountability measures or strategies for collaboration are shown to have positive outcomes in diverse policies, including for marine managed areas and natural solutions for adaptation and mitigation. Further, the policy options address a common set of drivers and pressures, as identified.

The appraisal provides broad indicative lessons about conditions for replication and achieving the policy goals. Simple attribution of positive outcomes to any one policy is problematic as multiple factors contribute to success. Weak systems for monitoring and tracking policy outcomes in social, environmental, economic and political domains mean that the appraisal relies primarily on qualitative analysis from peer-reviewed literature and documented project experience. Some of the identified policy options are in the early stages of implementation and consequently show limited impact; nevertheless, current results suggest potential for up-scaling and replication.

Transboundary Natural Resource Management

Transboundary approaches to environmental and natural resource management contribute to achieving the agreed goals in all thematic areas. There are many examples of successful transboundary initiatives throughout Africa, although there are significant variations in their focus, structure, delivery and scope.

These approaches demonstrate success in minimizing biodiversity loss, supporting integrated land and water management, improving local benefits, contributing to fairer and more equitable resource sharing including of ground- and surface water, and improving climate mitigation and the availability of resources for adaptation. Importantly, transboundary approaches often enhance cooperation and reduce confict by facilitating dialogue, establishing networks -including of marine protected areas - and encouraging learning and knowledge sharing. This helps create the political stability needed for economic and development cooperation. There are, however, many challenges.

Transboundary approaches are inherently complex processes involving many actors, issues and agendas. Strengthening dialogue can contribute to achieving consensus, but reaching it is nonetheless challenging. Eforts to bring different countries and sectors of society together can create rifts and alienate some communities, and scaled-up management may marginalize local users from decision making and reduce access to valuable livelihood resources. Implementation can also be impeded by ill-defned rights to land and resources, weak governance processes, and conficting interests and goals. Consequently, the development and harmonization of laws and policy are essential.

The rapid increase in transboundary natural resource management demonstrates that this policy, despite some challenges, has high potential for replication and for managing Africa's diverse shared ecosystems. For example, given that 75 per cent of African countries are coastal and that 70 per cent of river basins are shared by two or more countries, collaborative governance is essential for sustainable approaches.

Marine Managed Areas

Marine managed areas are part of a suite of approaches applied in Africa that contribute directly to achieving the oceans and seas goal, and also to the biodiversity and climate change goals by securing coastal ecosystems and environmental services.

The objectives of marine managed areas - which often include strictly protected no-take zones or other marine protected challenging. Consequently they are often smaller and wider apart than is ecologically viable.

Establishing managed marine areas can be an effective alternative, as they include multiple management zones and protected no-take zones. The information required for the design of such no-take zones can be obtained through rigorous quantitative research in a few representative sites combined with comprehensive surveys of traditional knowledge. Once designated, many managed marine areas face a lack of adequate resources for proper enforcement of regulations. However, alternative enforcement approaches can be used, including local community guards. An added advantage of managed marine areas is that they can control any unsustainable use that is displaced by fully protected no-take zones. Individually, however, countries may be unable to address this as use may occur in areas beyond their legislative jurisdiction, such as the high seas, all of which suggests a need for more collaborative and transboundary approaches. For example, Africa's widest network of managed marine areas, which stretches over 23 sites in six countries of Western Africa – Cape Verde, the Gambia, Guinea, Guinea Bissau, Mauritania and Senegal – has had considerable success in ensuring that fisheries, tourism, and oil and gas development do not adversely affect the marine ecosystem and its biological resources.

Africa's 45 649-km coastline, which encompasses 33 of the region's 48 mainland countries and areas in addition to multi-use areas – complement a broad range of national development and economic goals other than biodiversity conservation. These goals include improved food security, better livelihoods, effective governance and sustained economic growth. Managed areas complement other regulatory policies such as fisheries and water quality management. For example, five-year rotational harvesting in managed areas of the east coast of South Africa contributes to the rapid recovery of oyster populations during fallow years.

Although the establishment of marine protected areas has often been relied on to improve marine conservation, they face multiple challenges. Disparities in governance, institutional capacities, wealth distribution, social capital and availability of ecological data can affect both their establishment and their effectiveness. In some cases, marine protected areas face opposition from adversely impacted sectors of society. For example, tourism operators resisted the establishment of marine protected areas in Kenya because they could not aford the licensing fees, protective clothing, insurance and equipment required by new regulations. Local fishers who are excluded from their previous fishing zones may also oppose protected area development. In addition, many countries cannot aford comprehensive research on all marine habitats within their jurisdiction, making the identification and development of marine protected areas six island nations, illustrates the importance of strengthening marine management. Up-scaling managed areas and establishing a network is a step beyond the more traditional approach of establishing them opportunistically as single independent entities. Through interconnections and interdependencies, the individual elements of the network contribute positively to each other's integrity by decreasing overall vulnerability. Marine food webs extend beyond the boundaries of individual areas, and fishers are dependent on different species and geographic regions at different times of the year. Tourism revenues from an accessible managed area with charismatic species can help subsidize the maintenance costs of more remote places with no other values that can be easily captured through current market mechanisms. Many biophysical and socio-economic connections overlap national boundaries, and regional cooperation can promote national interests. Currently unmanaged areas merit priority attention within a larger managed area framework. As part of a regional marine conservation agenda, the formation of managed networks in those areas where coverage is minimal and urgently needed – including Northern Africa (Mediterranean Sea), northeast Africa (the Red Sea), the Gulf of Guinea and Southern Africa – has seen some initial successes.

Regional Approaches to Marine Pollution Management

Regional approaches that include a mix of self-regulated, state-enforced and collaborative management are effective in addressing the multiple drivers and diverse scales of marine pollution, and hence in achieving the selected goal for oceans and seas. A reduction in marine pollution also contributes to the biodiversity and climate change goals.

Coastal urban growth contributes to residential efuent, industrial discharges, storm-water run-of, agricultural and mining leaching, contaminated groundwater seepage, and industrial and vehicle exhaust fumes that enter the marine environment. The coastal cities of Accra in Ghana, Douala in Cameroon, Lagos and Port Harcourt in Nigeria and Luanda in Angola, for example, are all adversely affeed by industrial pollutants. Oil spillage and discharge from marine transport present major management and regulatory challenges, especially for oil-producing countries such as Libya and Nigeria, where problems are severe. Ofishore exploration, especially for oil, contributes to pollution from dumping at sea, accidental and intentional oil spills, engine leaks and noise.

Comprehensive regional marine pollution conventions govern the four major African coastal regions. The Convention for Co-operation in the Protection and Development of the Marine and Coastal Environment of the West and Central African Region (Abidjan Convention) and the Regional

Convention for the Conservation of the Red Sea and Gulf of Aden Environment (Jeddah Convention), and their associated protocols, provide important regulatory mechanisms for high-use areas and employ a self-regulatory approach. The benefit of self-regulation is that it is quick to respond, flexible, and sensitive to market circumstances. The primary drawback is that the responsibility falls on the industry to control pollution, and the incentives to do so may be insufcient. Waste exchange programmes, initiated under the Guinea Current Large Marine Ecosystem programme, have effectively supported waste reduction and ecosystem recovery. In Ghana, this has focused on using the waste from one industry as raw material for another.

Action plans to increase the capacity of port waste reception facilities have been developed regionally, although on-the-ground progress has been limited. For example, the Benguela Current Large Marine Ecosystem programme has promoted the sharing of facilities between ports in the Benguela and Guinea Current regions through assessment of reception facilities, technical training needs and regional capacity requirements as related to the International Convention for the Prevention of Pollution from Ships 1973, as modified by its 1978 Protocol (MARPOL 73/78). This process has increased the involvement of key regional stakeholders and advanced some key operational areas of the convention. The Port Management Association of West and Central Africa incorporates marine pollution compliance and requires further investment from ports and industry partners. The need to address the shortage of technical management capacity has been tackled by the International Ocean Institute– Southern Africa through targeted regional training for countries of Western Africa. The existing network of large marine ecosystem programmes, port management associations and regional conventions provides further opportunity for replication of this model.

Tax-based policies can complement this approach by extending a company's liability for environmental damage, although a common argument against tax-based policies is that they give the right to pollute if they are not punitive enough. When citizens have standing in the courts they can serve as an important check on industrial practice, as the Niger Delta Ogoni case demonstrates, a case that is also pertinent to the policy option on human rights. This in turn provides an incentive for improved environmental performance.

Although the regional conventions and their protocols are relatively comprehensive in addressing the various marine pollution issues, significant risks remain due to the lack of implementation of these regimes in some countries. But successes exist – as demonstrated by developments under the Convention for the Protection, Management and Development of the Marine and Coastal Environment of the Eastern African Region (Nairobi Convention).

With a network consisting of existing regional International Maritime Organization (IMO) ofces in Western and Eastern Africa, various regional conventions, and the regional seas and large marine ecosystem programmes, it is clear that the appropriate policy platforms are in place to combat marine pollution. However, there is a paucity of capacity for management in terms of equipment, technical training and institutional support, as well as for implementing existing policies, making investment in these areas a priority.

Ecosystem Services and Biodiversity Offsets

Innovative mechanisms such as payment for ecosystem services and biodiversity offsets contribute to achieving all the identified goals by encouraging, compensating and rewarding environmental custodians for maintaining or restoring valued environmental services.

A growing portfolio of payment for ecosystem services in Africa demonstrates benefits for both nature and people, including for watershed services in Eastern and Southern Africa. Biodiversity offset programmes have been adopted in Ghana, Guinea, Madagascar and South Africa. Payment for ecosystem services and offset approaches have also been used to support eco-labelling and community tourism, to protect fragile and valued habitats including forests, mangroves and coral reefs, and to sequester carbon (REDD+).

Despite some positive outcomes from these approaches, barriers to success remain. Opportunities for local communities continue to be limited: for example, large landowners or companies supply most biodiversity offsets even though low-income communities could be competitive suppliers of biodiversity compensation. The weak negotiating capacity of communities makes it difficult for them to participate and secure livelihood benefits that exceed their opportunity costs.

There is considerable potential for expanding payment for ecosystem services in Africa, as the region lags behind others in developing such approaches. In the global carbon offset market for 2011, for example, Africa accounted for less than 3 per cent of emission reduction projects, albeit the region has seen a strong growth trend in the past few years. Enabling factors include agreeing to a set of principles; strengthening the legal framework including certification and capacity building of buyers and sellers; encouraging participation by small-scale players by granting them land title or use, access or co-management rights; focusing on long-term livelihood assets rather than on short-term benefits; reducing corruption and "rent-seeking"; creating a more transparent business framework; and facilitating bi- and multilateral knowledge-sharing initiatives.

Reducing Emissions from Deforestation and Forest Degradation

Reducing Emissions from Deforestation and Forest Degradation (REDD+), including the role of conservation, sustainable management of forests and enhancement of forest carbon stocks, is a payment for ecosystem services mechanism currently being negotiated under the United Nations Framework Convention on Climate Change (UNFCCC). Various multilateral processes are underway to support countries in preparing for REDD+ implementation.

With the right safeguards in place, REDD+ could support climate mitigation through carbon sequestration – the climate change goal – and could also address important social and environmental dimensions that could lead to improved livelihoods. Depending on design, REDD+ initiatives may also ofer new incentives for addressing the biodiversity and freshwater goals by enhancing forests and the land goal by reducing economic reliance on land-degrading activities. If the current focus is extended beyond terrestrial forests to include mangroves, REDD+ equivalents could also support realization of the oceans and seas and land goals.

Although REDD+ is in its preparatory phase, there are some readiness activities, pilot projects and bi-lateral initiatives as well as carbon sequestration projects from which lessons can be garnered. Preliminary evidence suggests benefits for climate mitigation and the environment as well as for people, primarily through supplementary income.

Early lessons from carbon sequestration projects suggest that unless several challenges facing REDD+ are resolved, market-based approaches could fail to achieve positive outcomes, or might even increase global emissions. For success, REDD+ needs to address the enabling factors identified under the payment for ecosystem services policy option, and must also ensure that:

- earnings exceed opportunities foregone from agriculture and the fuelwood market;
- secure carbon rights that encourage equitable benefit distribution, reduce the potential for confict, and discourage forest conversion are adopted;
- enforceable social and environmental safeguards, such as free, prior and informed consent measures, are effective in reducing adverse impacts;
- systems for accurate measurement, monitoring and reporting of emissions are implemented;
- reduced implementation costs are achieved;
- effective intersectoral cooperation is established.

A potentially important limitation of REDD+ is that the current UNFCCC forest defnition excludes vast areas of open forests, generally in the dry tropics, and therefore overlooks important carbon stocks, for example in much of Eastern and Southern Africa where significant deforestation is taking place. The inclusion of these dry forests and woodlands would expand the relevance and impact of a post-Kyoto REDD+ mechanism.

As significant carbon stocks are held in coastal systems and soils, carbon credit schemes could be built into the design of new marine protected areas as public-private partnerships to enhance management and fnancing. Addressing the specifc circumstances of communities and organizations running these REDD+ initiatives is also important, as in coastal zones their investments may be at risk from natural disasters. A further challenge for REDD+ is that climate mitigation activities are poorly integrated with adaptation and development. This is particularly problematic given Africa's high levels of poverty and vulnerability to climate change.

Integrated Coastal Zone Management

Integrated coastal zone management provides a management framework that takes into account complex, non-linear interactions between and within human and ecological systems and across temporal and spatial domains, and consequently takes a significant step towards coherent management of entire ecosystems. It prioritizes the land-sea interface with the objective of balancing economic development and environmental protection, consequently contributing to all five selected goals.

The number of African coastal countries adopting integrated coastal zone management increased from five in 1993 to 13 in 2000; this is supported by specifc commitments to integrated management in regional agreements.

As a cross-sectoral approach, integrated coastal zone management involves all levels of governance and encourages the involvement of all stakeholders. This is well illustrated for Eastern Africa's coastal countries in the operation of the Secretariat for Eastern Africa Coastal Area Management and in South Africa. Success requires action and commitment at regional and national levels.

Experience demonstrates that the integrated coastal zone management protocol of the Barcelona Convention could be strengthened substantially through the use of spatial planning tools. Although these have been used terrestrially for decades, the broader marine community has only recently adopted them. New technologies, including remote sensing, geographic information systems (GIS) and spatial modelling, provide vastly improved capacity to replicate the spatial structure of nature in models of human-environment interactions, and support a strategic decision-making process that creates a blueprint for ocean use. A key strength of these technologies is that they explicitly recognize that there are valid competing demands on natural resources and that ecosystem-based management solutions must work within the capacity of the local communities. As a result, these tools encourage the development of equitable and viable solutions to the conservation of socio-ecological systems.

Sustainable Land Management

Sustainable land management can strengthen the management of water and land while incorporating social and economic values. Consequently, it supports achievement of the land and freshwater goals and contributes to the biodiversity, oceans and seas, and climate change goals.

An example of this approach is the TerrAfrica initiative. This multi-partner platform for consultation and action includes intergovernmental and civil society organizations. In partnership with the governments of Burkina Faso, Ghana, Namibia and Uganda, TerrAfrica supports country-level approaches. Dialogue on sustainable land management has been initiated in several countries including Eritrea, the Gambia, Malawi, Mali, Niger, Nigeria and Senegal. The success of TerrAfrica, including in Burkina Faso, Ethiopia, Ghana, Mozambique and Uganda, suggests a high potential for integrated and participatory land management approaches to be replicated in other countries.

Recognizing that addressing climate change as part of sustainable land management is essential to ensure adaptation and address climate-related land changes, TerrAfrica established climate change as a core priority in 2009. Adaptation policies should complement farmers' responses to climate change, including water harvesting and natural solutions such as ecosystem restoration.

Sustainable land management approaches appear to be most successful where there is high-level political support and when they build upon local knowledge and practice. This contributes to creating effective stakeholder coalitions and platforms, improving the development, management and dissemination of knowledge, and more effectively leveraging the investments required for sustainable management activities.

An on-going challenge in establishing sustainable management is land tenure insecurity. Many governments are rectifying this through land tenure reform: for example, Niger's Rural Code establishes a framework for protecting and revitalizing pastoralism, where previous policies favoured crop farmers. It has promoted the preservation of pastoral areas and protected herders' rights to collective use since 1993, including rights to move their livestock in search of water and pasture. In 2010 the Rural Code was modified to address outstanding ambiguities. For example, although the code established land commissions as representative organs in which all stakeholders participate, people still turn to religious and customary leaders first to resolve any land issue. A remaining challenge is to stop the encroachment of pastoral areas by croplands as farmers migrate north under demographic pressure. Mozambique's experience demonstrates that in replicating sustainable land management and land tenure, greater attention must be given to community empowerment and the capacity of state implementing agencies. Given the similarity in land-use systems across Africa, these approaches could be replicated in other countries.

Human Rights and Environmental Protection

Policy approaches that incorporate human rights contribute to the selected freshwater goal, the land goal through better recognition of local tenure, and the biodiversity and oceans and seas goals by holding decision makers accountable for decisions that adversely affect the environment. Importantly, these approaches support the achievement of the Millennium Development Goals (MDGs) at the same time as delivering benefits to the environment. Conversely, a lack of rights is often synonymous with high levels of vulnerability, as experience with external investments in land deals in Africa illustrates.

Human rights are important for protecting people and the environment when there are strong incentives for natural resource exploitation, as is the case in much of Africa. Governance rights, including participation and free prior informed consent, help ensure that local people's rights are taken into account. Human rights provide a benchmark for making sustainable choices and encouraging equitable and non-discriminatory outcomes. Once decisions have been made, litigation can provide a basis for their evaluation. In Nigeria, communities have used human rights law to oppose oil exploration that has adversely affeed agricultural land and biodiversity, including for example in the court case *Kenule Beeson Saro-Wiwa, President of the Movement for the Survival of the Ogoni People (MOSOP) and Eight Others,* unreported 1995. In 2002, the African Commission on Human and People's rights found that under the African Charter the Nigerian government has an obligation to protect the well-being of the Ogoni People (*Social and Economic Rights Centre v Nigeria).* Giving effect to this decision would limit the way in which oil exploration takes place and ensure protection of the environment, health and livelihoods.

Governance based on human rights may appear to be cumbersome, but it encourages rigour in decision making and ensures that multiple issues and values are taken into account.

Local, Inclusive and Participatory Approaches

Policies that reinforce local rights to participate in environmental management help to strengthen stewardship, contributing to the biodiversity, land, water, oceans and seas, and climate change goals. These approaches can be incorporated across different conservation policies, such as sustainable land management, integrated coastal zone management, and natural solutions for adaptation to and mitigation of climate change. They broaden the livelihood base for millions of people, for example through transboundary natural resource management, marine managed areas and REDD+, strengthen local resilience, including through policies on water harvesting and natural solutions, and encourage learning across levels.

Since the 1990s, there has been a rise in the number of countries using local, inclusive and participatory approaches, and growth in the extent of land under this type of management. For example, the percentage of forests under community tenure in Africa's ten most forested countries increased during 2002–2008 from 1.2 million hectares to 6.1 million hectares. Several countries including Cameroon, Ethiopia, Ghana, Kenya and Senegal have policies that recognize sacred sites. Sites conserved by indigenous peoples and local communities can be successful in strengthening ecosystem management and restoring and maintaining biodiversity, and complement state protected areas.

Significant barriers to successful implementation remain for many local and participatory approaches. The inadequate enforcement and implementation of local rights remains a challenge: for example, government authorities have often been slow in allocating community entitlement to forests designated as community forests. Conficts between local and state laws, as well as suspicion about communities' capacity to achieve sustainable management, affect government willingness to transfer authority. Better understanding of the multiple meanings and values attributed to forests by local communities can establish a basis for locally appropriate institutional arrangements. Other barriers include the limited use of markets due to insufcient fnance, poor information and technology fows, inadequate market links, and communities' inability to exploit economies of scale. Enhancing capacity and entitlements, as envisaged by the Convention on Biological Diversity (CBD), will be critical to improving environmental and social outcomes.

Water Harvesting

Water harvesting is used to collect run-of or foodwater for storage in the soil or in tanks so that it can be used for the production of crops, trees or fodder and for domestic use. Water harvesting therefore supports the realization of the climate change goal in strengthening adaptation by ensuring access to freshwater and by reducing the run-of impacts of extreme rainfall events in tropical, sub-tropical and dryland conditions; in addition it is appropriate for both rural and urban communities. Rainwater harvesting also contributes to achieving the freshwater and land goals and to the biodiversity goal through the restoration of water catchments.

The importance of this policy option is underpinned by climate change and the understanding that by 2020 some 75–250 million Africans will live in water-stressed areas, while an increase in extreme rainfall events will adversely affect soils and settlements, including cities. Across the Sahel, innovative rainwater harvesting has been applied to hundreds of thousands of hectares, enhancing agricultural productivity and reducing human susceptibility to climatic variability. In Mali, research has quantified the impacts of rainwater harvesting on crop yield increases and ground water recharge.

Establishing effective water harvesting can be challenging, with access to resources, labour and skills being limiting factors. Families may not be able to aford storage facilities that cater to household size. The returns from water-harvesting investments can be long term, so weak land security for smallholders, and particularly for women, may make them reluctant to invest in such technologies.

Nevertheless, the potential for rainwater harvesting is significant and can be replicated in many countries. Integrating water management into national adaptation planning can support the uptake of such technology by addressing legal and policy constraints and increasing community access to financial resources and skills. Several countries, including Togo, recognize water harvesting as a priority in their national adaptation programmes. Support for local knowledge, practice and innovation can empower communities to act and results in the diffusion of water harvesting through farmer-to-farmer learning.

Expanding opportunities for water harvesting can include the rehabilitation of degraded dams, restoration of watersheds, and conservation of existing forests that contribute to water provisioning. These strategies can improve year-round supply of water, soil conservation, and the expansion of livelihood activities including in the agro-pastoral sector.

Adaptation to and Mitigation of Climate Change

The restoration and maintenance of ecosystems can provide valuable resources for climate adaptation, disaster risk reduction and mitigation, and thus help achieve the climate change goal. By enhancing environmental goods and services, ecosystem restoration can also contribute to the realization of the land, oceans and seas, water, and biodiversity goals.

Restoration can involve diverse actors at transboundary, national or community level and includes the maintenance of protected areas. By restoring or maintaining ecosystems, natural solutions provide opportunities for adaptation and mitigation. Mangrove restoration, for example, can enhance coping capacity by stabilizing coastlines. Mangrove restoration also supports adaptation through the provisioning of environmental goods such as food, fuel and wood. For example, Nigeria's mangrove forests provide breeding grounds for more than 60 per cent of the fish caught between the Gulf of Guinea and Angola. In Sudan, the restoration of rangelands – achieved through rotational grazing and a shift in livestock composition – helped improve livestock pasture and food security. An unexpected consequence of this effort was that pastoral nomads were attracted to the area, with confict being avoided by using traditional local institutions and values to negotiate access. Protected areas, including in Niger, have been found to support the *in situ* conservation of wild crop relatives, which are often more drought resistant than domesticated crops and can be used to strengthen agriculture and food security.

Ecosystem restoration frequently requires a coherent but cross-cutting multisectoral approach as drivers and pressures exist at multiple levels. Large-scale or global drivers include oil exploration, agricultural expansion and pollution, infrastructure and transport development, population growth and settlements, and coastal development. At the same time, local livelihoods can place pressure on resources where governance and management are weak, for example through unsustainable fuelwood harvesting in mangroves. Establishing integrated approaches that address drivers at multiple levels is often challenging, especially where coordination and collaboration between policy development agencies and policy implementing agencies are weak. Poor data collection, monitoring and information further constrain adaptive management. Inadequate legislation that is sector-based, conficting, defcient and unenforceable provides a weak basis for planning and management.

In addition, enhancing the conservation of ecosystems and their capacity to regenerate requires a better understanding of the links between different ecosystem components as well as of social-ecological resilience. Investing in and generating ecological knowledge and translating it into information that can be used in governance and policy development is essential for management success, and requires a better interface between science, policy makers and communities. Regional cooperation, community-driven strategies and public-private partnerships can support learning, improve sustainability and encourage ecosystem approaches. The recently adopted Mangrove Charter for West Africa, which is complemented by country-specifc action plans, is an example of this.

Given that adaptation is about local capacity, it is important that strategies and projects enjoy shared understanding between policy makers, technical agencies and communities. Unless this is achieved, there is a risk that adaptation strategies will run counter to local livelihoods, values and cultures, and that uptake will be low, as in the government-initiated resettlement scheme following Cyclone Eline in 2000 in Mozambique. The loss of easy access to resources and social support were key obstacles to support for resettlement. A second major challenge was the conficting perception of climate risk severity held by the government and the communities. These results highlight the need for active dialogue across stakeholder groups as a necessary condition for formulating and successfully implementing policies. Ongoing dialogue creates the basis for reassessing strategies and responding to change

Stakeholder Pollution Management

Pollution management is important for restoring ecosystems and realizing human health goals. It contributes to achieving the social and environmental aspects of the selected goals for biodiversity, freshwater, oceans and seas, and climate.

Africa has relied primarily on regulatory approaches to achieve pollution targets. These approaches influence environmental outcomes by regulating processes or products, limiting the discharge of specified pollutants, and restricting certain polluting activities to specifc times or areas. However, regulatory instruments are often inefcient in achieving pollution control objectives, especially where resources for monitoring pollution and compliance are lacking. The level of expenditure required for ensuring compliance with increasingly stringent environmental laws is an unmanageable cost for many governments. In contrast, stakeholder-driven management approaches have the potential to make pollution control economically advantageous to commercial organizations. These approaches can involve varying degrees of incentives, information and administrative capacity for effective implementation and enforcement. The principal types of economic instruments used for controlling pollution include pricing, pollution charging and marketable permits.

Building on Success

This environmental policy appraisal suggests that opportunities for building on existing success can be effectively harnessed to ensure better implementation and positive outcomes for people and the environment.

Replicating and up-scaling effective approaches is important, but policies should not be blindly replicated, and should be modified to achieve good ft with local, national and regional conditions. As amply demonstrated in the policy options detailed above, it is important to maximize opportunities by focusing on options that are mutually reinforcing and cross-cutting. Finding and developing synergies is cost effective when financial and human resources are limited. Ensuring that policies are not in confict with each other and that they do not lead to an externalization of adverse impacts is important.

As the policy appraisal shows, effective policy implementation requires reducing or removing barriers and strengthening enabling conditions. Insufcient monitoring, special-interest decision making, weak governance and rights, and a lack of adequate capacity have undermined policy success.

Policies that have in-built flexibility are needed to address environmental change. Investing in monitoring and evaluation, as well as social learning, supports revision and modification of policy responses, as illustrated in many of the policy options discussed here, including, for example, natural solutions for adaptation and mitigation.

Decision making that is strategic and takes account of how changes in environmental use and governance affect the resilience of the social-ecological system has been shown to be effective in securing economic, social and environmental benefits. Integrating human and ecological understanding and priorities in environmental management can help ensure that choices do not destroy or undermine the environmental resources that underpin future options. Such approaches – including ecosystem-based management – prioritize the interface between people and nature and do not favour just one ecosystem component, industry sector, community or socio-economic group. Ecosystem-based management is one way to maintain the Earth System's ability to adapt to change, as compared to other approaches that focus on fxed targets and state systems, or on hard-engineering solutions that often interfere with natural processes.

Strong accountability helps secure government and private-sector commitment to implementation and to achieve agreed outcomes. For countries to be better able to demonstrate results, systems for monitoring progress need to be established. Developing performance indicators rather than effort-based indicators, such as the number of meetings held, improves clarity about how and to what extent the purpose of the policy is being achieved. Strong and effective national and sub-regional reporting systems help hold implementing agencies to account and provide an opportunity to document successes, which in turn set the basis for up-scaling and replication.

Cooperation has been shown to be effective for achieving sustainable management, including policy options for transboundary coastal and land-based resource management, and where there are multiple stakeholders. This has improved equity, enhanced skills sharing and reduced confict. In some cases, external support and collaboration with donors have helped establish effective platforms for engagement, learning and sharing knowledge and skills, including in TerrAfrica and under the Nairobi Convention. Partnerships with the private sector and environmental managers or custodians have been shown to be effective in securing benefits in many of the policy options, including payment for ecosystem services, and in mangrove restoration. Several of the options presented, including sustainable land management, show that a high degree of participation at local and government levels helps to ensure relevance, with good outcomes for strengthening sustainability. Decentralization and devolution policies, including in community-based resource management, have achieved positive outcomes for communities and for the environment.

Strengthening the governance and institutional regime for more equitable benefit sharing is critical, given that ecological and social resilience are tightly inter-twined, as South Africa's basic water policy shows. Weak tenure and entitlements stand out as key barriers to achieving equitable benefits for payment for ecosystem services including REDD+, community-based management and other policy options. While these are national problems, the scale and commonality of the challenges suggest that developing and adopting regional or global protocols for cooperation and sharing could provide the basis for more effective engagement and management of benefits and losses. Strengthening and integrating human rights perspectives in environmental management frameworks at national and regional levels supports more inclusive, long-term approaches by protecting livelihood rights, ensuring inclusion and reducing confict. Regional human rights bodies can play an important role in solidifying the piecemeal environmental benefits that human rights recognition has already brought, especially where the mandate of regional courts and the rights of citizens to bring actions are strengthened, as shown in the human rights policy option.

Environmental policy is often out of step with realities on the ground, with governments acting alone often unable to effect the necessary change. Innovative institutional arrangements for pooling financial resources, knowledge and capacity, however, can contribute to achieving environmental goals. Improving capacity and equity among diverse communities, including governments, is essential to support collaboration and to secure rights. The policy options demonstrate the potential of various strategies to enhance capacity. At regional and sub-regional levels, for example, mechanisms for sharing information and knowledge, as in the management of marine pollution, could be better utilized.

References

Abdulla, A., Gomei, M., Maison, E. and Piante, C. (2008). *Status of Marine Protected Areas in the Mediterranean Sea*. IUCN, Malaga and WWF, France

Abdulla, A., Gomei, M., Maison, E. and Piante, C. (2008). *Status of Marine Protected Areas in the Mediterranean Sea*. IUCN, Malaga and WWF, France

FAO (2010). *Global Forest Resources Assessment.* Food and Agriculture Organization of the United Nations, Rome

Kanji, N., Toulmin, C., Mitlin, D., Cotula, L., Taoli, C. and Hesse, C. (2006). *Innovation in Securing Land Rights in Africa: Lessons from Experience*. International Institute for Environment and Development, London

Nelson, F. (2010). *Community Rights, Conservation and Contested Land. The Politics of Natural Resource Governance in Africa*. Earthscan, London

Stalk, A. (2004). *Management of the Free Basic Water Policy in South Africa*. Master project. Roskilde University, Roskilde

9

Management of Environmental Problems in Asia and Pacific

The global drivers identified in and particular unsustainable economic growth, population increase, mass consumption and urbanization - pose clear challenges to sustainable development in Asia and the Pacifc. It is therefore important that policy responses are designed to enable the best possible adaptation to the pressures and impacts deriving from these drivers.

Environmental Issues

The priority concern for most countries in the region is how to build resilience, especially in the most vulnerable communities, to climate change impacts already set in motion by past greenhouse gas emissions. Parts of low-lying Pacifc island countries may disappear entirely due to sea level rise, extreme weather events are likely to become more frequent, and marine habitats such as coral reefs and mangroves are threatened by increased temperature and ocean acidification.

Under a business-as-usual scenario, the region will contribute approximately 45 per cent of global energy-related carbon dioxide (CO_2) emissions by 2030 and, by one estimate, more than 60 per cent of total global CO_2 emissions by 2100. However, intra-regional diversity is great - China is the world's largest CO_2 emitter while most Pacifc island nations are among the smallest. Transport-related emissions are expected to increase by 57 per cent worldwide between 2005 and 2030, with China and India accounting for more than half of that increase. Nonetheless, there are encouraging signs on mitigation. At least ten countries in the region have voluntarily pledged greenhouse gas emission reductions, including Indonesia's promise of a 26 per cent CO_2 reduction compared to business-as-usual by 2020, and China's of a 40-45 per cent reduction in CO_2 per unit of gross domestic product (GDP) compared to the 2005 level by 2020. With one of the world's greatest potentials for mitigating CO_2 emissions being reduced deforestation alongside improved land-use management, Asia and the Pacifc can make significant contributions to global efforts at climate change mitigation. Accessing climate funds to enable these contributions, however, is a major concern for developing countries in the region.

Although UNFCCC Article 3 was selected by the regional consultation, three other goals (UNFCCC Article 2, the Bali Action Plan and the Delhi Declaration) selected by the GEO High-Level Intergovernmental Advisory Panel were also considered because adaptation, mitigation, capacity building and fnancing need to be considered as an integrated package of policy measures.

The imminent threat of mass extinction of species, brought about by continuing habitat fragmentation, degradation and loss, overexploitation of resources, invasive alien species, illegal wildlife trade, pollution and climate change are priority environmental concerns in the Asia and Pacifc region. The *Global Biodiversity Outlook 3* concluded that the 2010 goal of reversing biodiversity loss had not been achieved. The Strategic Plan for Biodiversity for 2011–2020 embodying the Aichi Biodiversity Targets now provides the general framework for biodiversity conservation.

The key environmental priorities in the water sector being faced by the region are the quantity and quality of water resources, climate change, access to safe drinking water and transboundary issues. All of these key challenges are reffeed in the selected goal.

The regional consultation also noted that JPOI Paragraphs 25d and 7a should be included in the assessment, as well as taking an innovative approach to links with other themes.

The chemicals and waste theme encompasses a range of interrelated issues, including the production and use of chemicals, hazardous waste, electronic waste, transboundary movement, product reuse, materials recycling and municipal waste management. During the regional consultations, JPOI Paragraph 23 was selected as the overarching goal for the theme, although JPOI Paragraph 22 was considered equally relevant.

The selected global goals are built around the concept of life-cycle thinking. Hence the starting point for effective policies is to use demand management and resource efciency to minimize waste generation and the use of hazardous chemicals. The region's political recognition of the need to prioritize waste minimization and resource efciency is not matched by policy implementation. Only weak efforts have been made to address the escalating use of resources and hazardous substances that eventually end up as waste and pollutants.

Environmental governance functions through institutions, laws, norms and processes for collective decision making, and the region has a wide diversity of systems and mechanisms. However, many "remain centralized, expert-driven, compartmentalized, and inflexible". A persistent problem is that, "many environmental laws, regulations, action plans and programmes [have not been] effectively implemented", making greater progress necessary to achieve the selected global goal of good governance at local, national, regional and global levels

Policy Analysis

The first step in the policy analysis framework was to formulate a long list of policy options with the potential to accelerate achievement of the selected global goals, and then identify a few priority policies or policy clusters for further analysis.

In some cases, the long list of policy options was clustered into groups of policies with a common intent prior to screening, for ease of assessment and acknowledging that most policies are implemented as part of a complementary package rather than alone. Here, priority means that the policy or cluster of policies was selected for more detailed policy analysis rather than implying high priority for a specifc country or sub-region.

The analysis of limitations is included because even successful policies may have side effes or unintended consequences that need to be understood and addressed during implementation and that may impede replication elsewhere. To illustrate how policy packages may be introduced in coordinated steps, a series of graphic representations show:

- possible time frames: short term, 1-5 years; medium term, 6-15 years; long term, 16 years or more; and

- direct policy measures that are intended to target the immediate cause of the problem, ranging to indirect policy measures that help to achieve the selected goals by tackling related issues.

Climate Change

The key element of the selected global goal for climate change is to take a precautionary approach to anticipate, prevent or minimize the causes of climate change and to mitigate its adverse effes.

The clean energy policy cluster includes a renewable energy mandate and potentially carbon capture and storage, which, if and when the technology is proven, could contain the greenhouse gas emissions from the largest source – globally 12 billion tonnes of CO_2 per year by 2020. This cluster also has significant co-benefits such as improved air quality and health improvements, avoidance of environmental damage from mining and exploration for fossil fuels, improved energy security and new green job opportunities, and may ofer households and businesses opportunities to generate their own energy and supply the surplus to the grid. Potential limitations include non-climate-related negative environmental impacts such as mining for rare earth metals; the competition between biofuels and food production and impacts on biodiversity; higher costs to end users; unproven technologies such as carbon capture and storage; and the impacts on businesses and employees in traditional fossil fuel energy production.

The energy efciency policy cluster is aimed at reducing energy demand through targeted efciency improvements in buildings, transport and agriculture, globally representing 14 billion tonnes of CO_2 per year by 2030. The principal benefits are reduced operating and travel costs, health benefits from cleaner air, reduced trafc congestion following the growth of mass transit systems, and lower environmental impacts from low-tillage and other energy-saving agricultural practices. Limitations include the high initial costs of retroftting buildings and introducing or expanding mass transit systems, high resettlement costs when people are displaced by the building of above-ground mass transit systems, and the potential lack of uptake of mass transit by car-owning urban households. Furthermore, energy efciency improvements such as hybrid cars, while desirable, may be only temporary or partial due to widespread rebound effes whereby part of the efciency gain is offset by increased energy use. Inadequate public awareness and incomplete markets for energy efciency are also barriers to greater market penetration.

The technology policy cluster includes policies that promote technology transfer and diffusion such as technology transfer agreements, intellectual property rights waivers or buy-outs, and domestic research, all of which will also contribute to achieving a precautionary approach to combating climate change. This policy cluster will allow developing countries to leapfrog development stages and avoid the carbon-intensive trajectory of developed countries. It will improve human well-being, assist national development budgets by avoiding a lock-in to fossil-fuel-intensive approaches, and build national capacity for research and development. As a political strategy, such an approach has few limitations, although some intellectual property holders may be resistant to waivers to maintain their global competitiveness, and may be disadvantaged by being pressured to release their rights at less than market value. Indeed, some research claims that stronger intellectual property rights increase international technology transfer. Technology transfer from developed to developing countries may not be effective if the latter have inadequate domestic capacity to apply the technology in terms of infrastructure, human and financial capital, and the appropriate institutional environment.

The financial policies cluster includes greenhouse gas emissions trading, the Kyoto Protocol's Clean Development Mechanism, the Joint Implementation mechanism and the Adaptation Fund, amended tarifs and taxes such as feed-in tarifs, a carbon tax and an aviation tax, or elimination of subsidies that promote fossil fuel use and of financial incentives that encourage inappropriate land use and forest loss. A carbon tax may allow governments to reduce other taxes or generate additional revenue streams to invest in sustainable development. Earmarked fnancing provides incentives for scaled up investment in low-carbon technologies by the private sector. Penalizing fossil-fuel-based industries through taxes and tarifs levels the playing feld for emerging technologies. Potential limitations include intermediaries benefiting from carbon market initiatives rather than achieving low-carbon objectives. Global schemes such as the Clean Development Mechanism tend to be bureaucratic and cumbersome, and to benefit too few countries. Social impacts, particularly for the poor, may be high due to more costly goods and services unless these are offset by lifeline support measures or tax rebates. The distributional impacts of all financial measures need to be carefully analysed prior to adoption.

The adaptation policy cluster anticipates the likely impacts already embedded in the climate system due to historical levels of greenhouse gas emissions and ensures that communities can adapt to inevitable changes. Policies that facilitate adaptation include mandatory infrastructure design for future climates, or climate proofng; ecosystem-based adaptation in planning and zoning schemes; building climate resilience in agriculture, forestry and fisheries; and integrating climate change adaptation and disaster risk reduction. These policies have a range of co-benefits such as biodiversity conservation, improved recreational opportunities and managed natural resource harvesting. The main benefits are a lower incidence of death or injury from extreme storm events and droughts; reduced future economic and social costs; additional economic opportunities for the construction sector; higher property values in secure areas; and increased security and resilience of affeed communities. Limitations include the environmental costs associated with large-scale infrastructure such as higher food levees; social costs if communities or infrastructure need to be relocated away from vulnerable zones; investment costs in climate proofng and possible compensation costs for affeed properties and companies; and the political costs of diverting funds to retroft old and new infrastructure.

The policy cluster on land management for carbon sequestration aims at reducing greenhouse gas emissions from unsustainable land-use practices, including forest loss, biomass decomposition after harvesting, peat fres and decaying drained peat soils, which may contribute 15–20 per cent of total global emissions. In South East Asia, emissions from inappropriate land use and forest loss may account for as much as 75 per cent of the total from that sub-region, mostly from forest loss in Indonesia. Halving deforestation rates by 2050 and maintaining that level until 2100 would account for 12 per cent of the total emission reductions needed to stabilize atmospheric CO_2 at 450 ppm. Protection of coastal wetlands and marine ecosystems can also mitigate emissions. Principal benefits include the conservation and supply of ecosystem services such as biodiversity and water supply and quality; maintenance of indigenous cultural practices; soil conservation; and promotion of local livelihoods. Limitations include possible confcts with other development objectives; impingement of local economic aspirations due to restrictions applied by protected area managers; and more costly land management practices.

Biodiversity

The selected biodiversity goal contains elements of conservation of biological diversity, sustainable use of its components, and the fair and equitable sharing of the benefits of using genetic resources.

The policy cluster for conservation of biodiversity promotes the creation of protected areas, including areas that connect landscapes and seascapes, through identifying areas of high but threatened biodiversity value and biodiversity corridors that link protected areas as a system. The notable progress in the establishment of terrestrial and marine protected areas reported in the *Global Biodiversity Outlook 3* may be attributed to explicit policies on protected areas, with many countries in Asia and the Pacifc using legislation to establish protected areas. Existing policies on protected areas may need further improvement, yet they provide a good foundation for attaining the global objective of biodiversity conservation. Commitments like the Convention on Biological Diversity (CBD), the Ramsar Convention on Wetlands, the World Heritage Convention and new funding mechanisms often drive both the establishment and improved effectiveness of protected areas. A shift from revenue generation to a conservation policy mandate has effectively reduced associated illegal land-use change.

Challenges remain in ensuring that protected areas are part of an ecologically representative network, on land and at sea, and to provide effective protection for threatened and endemic species. Many protected areas in Asia and the Pacifc have communities within or on their periphery for whom formal recognition of stewardship, traditional livelihoods and conservation traditions is necessary. Formal recognition of indigenous and local community-conserved areas would increase the ecological coverage of legally protected areas and support community rights in protecting the sites. Possible limitations include competition with development objectives, difficulties in measuring the real values of conservation against development activities yielding shorter-term economic returns, and institutional and individual capacity to effectively enforce conservation laws.

Also covered under this policy cluster are regional cooperation initiatives on transboundary protected areas and biodiversity corridors. Transboundary collaboration fosters the cooperation of national institutions to benefit multiple countries, as demonstrated by several examples involving cross-boundary interest in protecting areas with high levels of biodiversity such as the Greater Mekong sub-region, Terai Arc landscape in India and Nepal, Sulu-Sulawesi marine areas and the Coral Triangle. The benefits from this cooperation are an augmentation of national efforts, transfer of capacities across countries and twin conservation efforts involving several stakeholders across borders. The key challenges are sustainability, difering capacities of institutions involved and the political nature of the cooperation whenever sensitive sovereign issues arise.

The targeted species conservation policy cluster is intended to protect species such as tigers, elephants, pandas, saola – an extremely rare antelope discovered in Viet Nam and Laos in 1994 – or other species of biological, economic, spiritual and cultural importance, including species covered by the Convention on International Trade in Endangered Species of Wild Fauna and Flora (CITES). These policies protect species from capture and hunting, being kept as pets or being traded for medicinal or food purposes, as well as promoting their well-being in captivity, such as in wildlife parks or zoos. The policies not only protect valuable species but also help to spread a broader conservation message of the need to conserve species and their habitats. Charismatic species also act as rallying points for political support, promote nature-based tourism and help leverage resources for broader institutional support. The main limitations are that there may be too much attention on such individual species, drawing support away from others in their native habitat.

Invasive alien species, especially in island ecosystems with high endemism, are a major threat to species conservation. Apart from national quarantine processes and regional networks like the Asia-Pacifc Forest Invasive Species Network, control mechanisms are limited. However, there have been a few cases of successful eradication of invasive alien species in the region.

The illegal wildlife trade policy cluster, based around CITES, is intended to eliminate the illegal wildlife trade, including by strengthening border controls and training customs ofcials to recognize endangered species, and through campaigns to raise awareness and publicity surrounding successful enforcement cases. Benefits include the protection of natural assets and saving certain species from extinction. There are also significant co-benefits in the form of improving law and order generally (as wildlife crime is often associated with other criminal activities), improved governance structures and better response mechanisms for wildlife protection within national borders. The main limitations are the need for heavy investment in law enforcement, possibly denying the customary rights of indigenous communities to access non-timber forest products, and possible social confict where bushmeat is an accepted social norm and a significant source of protein.

The community management policy cluster includes co-management, stewardship of traditional owners, recognition of property-use rights, and policies establishing various sustainable forest and fisheries management schemes. Benefts include better conservation outcomes, expanded livelihood opportunities, income diversification, poverty reduction, reduced tension between state and citizen, and better governance and institutional reform. Potential limitations are the capture of economic benefits by an elite, exclusion and conficts within communities, and a need for both long-term investment and capacity building.

The policy cluster on innovative fnancing mechanisms provides incentives for communities to remain engaged in conservation and institutionalizes their involvement. Countries testing innovative fnancing mechanisms like payment for ecosystem services under a green growth paradigm are increasing. Nature-based solutions to climate change – such as Reducing Emissions from Deforestation and Forest Degradation (REDD+) – ofer potential for social and biodiversity co-benefits by avoiding land-use change, which is the major driver of biodiversity loss in the tropics. Challenges to mainstreaming economic instruments for conservation remain in terms of high resource degradation and increasing pressure, as well as difficulties in streamlining supportive policy and legal regimes, institutional mechanisms, and equity and rights of the community.

The policy cluster on access and benefit sharing for the equitable use of genetic resources includes recognition of the rights of indigenous stewards of ecosystems, intellectual property rights protection and regulations preventing biopiracy. The policy cluster draws heavily on the outcome of CBD negotiations on access and benefit sharing, notably the Bonn Guidelines and the subsequent international regime. Adoption of the Nagoya Protocol on Access to Genetic Resources and the Fair and Equitable Sharing of Benefts Arising from their Utilization will guide the ongoing effort for developing national and regional agreements. The Association of Southeast Asian Nations (ASEAN) draft agreement on Access to Biological and Genetic Resources, combined with draft policies and laws on access to genetic resources, benefit sharing and traditional knowledge in Bangladesh, Cambodia, Mongolia, Nepal and Sri Lanka, will provide additional incentives for implementation.

The principal benefits of these policies include providing an additional incentive for indigenous communities dependent on natural resources to maintain a full range of biological diversity, a fair and equitable return from investors that stand to make a profit from capitalizing on indigenous knowledge and/or stewardship, and a means for governments to protect a national heritage. The main limitations are the difficulties in identifying communities practising traditional knowledge, and its objective validation both for effective rewards and for overcoming potential constraints on voluntary research.

Water Resources Management

The global goal on freshwater selected for Asia and the Pacifc targets improvement of water allocation, conservation of both the quantity and quality of water resources, and the safeguarding of ecosystems. These could be achieved by systematic promotion and application of integrated water resources management as a policy-planning framework.

The concept of integrated water resources management has been in place for almost 30 years and many countries have tried to set up institutional frameworks for it. In practice, however, only a handful of countries in the region have established the necessary legal and institutional capacities for its implementation. In most countries, organizational reform has involved setting up apex water bodies and river basin organizations to implement the integrated approach. However, water resources are still managed primarily through a sectoral approach in which policy is shaped by each agency in charge, often without cross-sectoral coordination and with occasional tensions. The application of integrated management can contribute to sustainable water resource use by integrating and balancing human needs with conservation and the restoration of ecosystems, and will enable the region to cope with complex and unpredictable challenges such as future climate change impacts related to changing patterns of extreme events including droughts and tropical cyclones (typhoons).

The water allocation and cooperation policy cluster aims to achieve an acceptable balance between potentially competing water uses and involves enhanced confict management and cooperation capacities. Typical policies may involve assigning existing local communities to implement water resources management, establishing water users' associations or river basin organizations with explicit mandates to manage potential conficts, using multi-stakeholder participation to draw up basin development plans, or setting up confict resolution panels. Cooperation institutions established in river basins have helped to reduce potential water related conficts in Central Asia. For example, the inclusion of water users in the canal water management system in the Fergana Valley of Central Asia improved transparency and more equitable water distribution. In Australia, the Murray-Darling agreement and its implementing agency is another example of basin-level institutional arrangements that have been emulated in other countries.

The pioneering of community-based participatory irrigation management in Andhra Pradesh, India shows the importance of water users' associations in leading such initiatives. While the policy has not been an unqualified success, a state-level federated structure of users' associations was created, followed by district- and farmer-level organization structures with well-defned mandates, transparent systems and people-centred decision making. At the local scale, these policies have contributed to increased financial capacity through revenue generation from the group members and the incorporation of a watershed management approach. The main limitations have been difficulties in scaling up from a pilot project level due to resistance from those who were benefiting from the *status quo*, prolonged water scarcity, and unsustainable institutional arrangements once external project funding came to an end.

The basic human needs policy cluster relates to satisfying the need for water and is oriented towards improving and/or increasing the volume of stored water along with the resulting flexibility in use. Policies include planning guidelines to store urban storm water run-of for watering public gardens or for street cleaning, or requiring buildings above a certain size to store water collected on the roof. Rainwater harvesting is already included in national and local policies in some countries such as Sri Lanka, Australia and in some states of India. Support for construction of farm dams and renovating existing dams to extract greater volumes of water through application of new technologies are feasible policy options. National policies on the construction of large water storage facilities for food control, hydropower and irrigation have potential immediate benefits, but often induce concerns over environmental and social impacts. The water-use efciency policy cluster promotes the application of economic instruments, especially in urban areas, including graduated pricing based on the type of water use and volume, pollution charges and combined payments for water and sanitation. The benefits of economic instruments include behavioural change inducing water saving by users, increased revenue for social measures and the operation and maintenance of the water supply. Experience from Cambodia and the Philippines shows that it is possible to improve access to safe drinking water significantly in cities and rural areas. The principal limitations are the difficulties of making water a priced service rather than a free good, higher management costs, potential cost burdens on some users, shifts to non-priced water resources such as groundwater, unpopularity with politicians who are reluctant to impose new charges and delays in effective institutionalization. Reducing the use of un-priced water is a way of improving water-use efciency but it needs investment and staf capacity development. Manila Water and the Phnom Penh Water Supply Authority have significantly reduced un-priced water use through management improvements.

The water environment policy cluster includes strengthening legislation and the implementation of water quality management by ensuring regular monitoring and reporting, applying the polluter-pays principle through pollution fees, incorporating total pollution load control systems, and mandating treatment levels for different categories of wastewater. The promotion of industrial and domestic wastewater treatment also contributes to improving water quality. Public communication strategies such as a water quality index should also be considered as a way of drawing attention to changes in water quality. The main benefit is that sufcient water of the desired quality becomes available, thus reversing current trends. Improved water quality reduces health risks and helps achieve one of the Millennium

Development Goals. The benefits to industry and the water supply sector are lower water treatment costs. Limitations include the high wastewater treatment costs and the difficulty in convincing polluters, especially marginal small to medium-sized enterprises, to undertake the necessary control measures voluntarily, thus requiring resolute implementation of the polluter-pays principle and command-and-control regulations. Ensuring environmental fow, which requires a high level of political commitment, is also important for the health of water bodies.

Chemicals and Waste

The global goal for chemicals and waste focuses on life-cycle analysis, transparency and risk assessment for a participatory approach to minimize risks to human health and the environment.

The Asia and Pacifc region is facing rapidly growing challenges in waste and chemicals management, fuelled by the region's combination of strong economic growth and population increase, and its rapid industrialization and urbanization. In low- and middle-income countries in particular, volumes of waste are growing and waste streams are becoming increasingly complex, containing ever larger amounts of hazardous substances. The capacity to collect and properly treat this waste is lagging, with consequent impacts on human health and the environment. Similarly, the use of agricultural and industrial chemicals and the unintended generation of hazardous substances are escalating, leading to impacts that are currently poorly monitored and therefore little understood.

The policy cluster on product redesign and sustainable consumption addresses these problems at source, but it will take a relatively long time to realize its full potential. Efective solutions to the challenges of waste and chemicals require preventive approaches. Substances with hazardous properties need to be phased out and replaced with safer options as far as possible. Production systems and products need to be redesigned with their full life cycle in mind, minimizing resource consumption, chemical hazards and waste generation. Overall, there is a need to encourage more sustainable patterns of consumption that can deliver quality of life with the minimum environmental burden.

Several actions towards this objective are already being taken. Green public procurement has proven an effective tool for creating a market for improved products, including those made of recycled materials. Extended producer responsibility for electronics has promoted design changes and fostered closer collaboration between manufacturers and recyclers. A good example of chemical substitution has been the phasing out of ozone-depleting substances under the Montreal Protocol on Substances that Deplete the Ozone Layer. Waste avoidance has had some success for industrial waste but has proven more challenging for household waste.

The principal benefits of such policies include minimizing the need for waste treatment and chemical safety measures, and a reduction in both illegal waste dumping and pollution. The main limitation is that some companies may have to internalize costs that they were previously able to shift to the public domain, although evidence suggests that taking a life-cycle approach can often also reduce production costs.

The policy cluster on collection systems and treatment facilities aims to ensure that, once products are in circulation, priority is given to reuse and recycling, with energy recovery and safe disposal as less desirable options. Industrial and municipal solid wastes pose different challenges and require different approaches, although there may be synergies in combined treatment. Efective policy interventions need to cover all stages, including efcient collection systems, safe treatment facilities and functioning markets for recovered materials. The cost of establishing systems of a high environmental standard may seem challenging, but the cost of inaction is also significant. Efective technologies are widely available, including both high-tech solutions and systems based on traditional practices. In Japan, for example, more than 50 million tonnes of solid waste are burned every year, often generating electricity and providing district heating. In 2000, Japan set a target to reduce dioxin emissions from incinerators by about 92 per cent from 1997 levels – a target achieved by 2003, with over 2 000 industrial waste incinerators shut down. The revised target in 2003 was to achieve a further reduction of 30 per cent by 2010. The Republic of Korea has been particularly successful in introducing source separation of food waste for separate treatment, a policy that has significantly reduced the emissions of greenhouse gases from waste treatment. Community-based composting has also been successfully introduced in many cities across Asia and the Pacifc, reducing disposal needs and the associated costs for municipalities.

The principal benefits of these policies include the obvious direct environmental and health benefits as well as the longer-term indirect and/or cumulative impacts of lower concentrations of toxic and hazardous materials. The main limitations of recycling have been associated with finding markets for recovered materials, leading to diminished private-sector enthusiasm after an initial period.

The international collaboration policy cluster addresses the need for a joint approach to meeting challenges as the region's countries become more integrated. This is particularly relevant to waste and chemicals, since end-of-life products and recyclable materials are increasingly traded across borders, and many chemical pollutants, such as persistent organic pollutants

(POPs) and mercury, can spread far from their sources. Developing countries and economies in transition are facing particular challenges in formulating and implementing effective policies to address these problems: international collaboration can provide the technical and financial resources needed. Regional monitoring systems for chemicals in the environment such as the POPs Global Monitoring Programme have been instrumental in identifying pollution sources for the formulation of effective action.

The benefits of these policies depend on the form of collaboration. Strengthening the technical and governance capacity for waste and chemicals has obvious and direct benefits for human health and the environment. Efective and safe treatment of hazardous materials can be ensured if these operations are carried out in countries with adequate capacity and if shipments to other countries are prevented. Enhanced international information sharing enables traceability and effective preventive action, including strengthened control of transboundary movements.

Environmental Governance

The global goal on governance relates to strengthening the multiple dimensions of sustainable development. Four key policy clusters were identified that could accelerate its achievement. Capacity development, access to education and information remain as underlying enabling factors for the effectiveness of each policy cluster.

The policy integration and mainstreaming cluster aims to integrate sustainable development functions, which are commonly fragmented between different ministries and agencies with weak coordination. Increased policy integration and strengthening the capacity of environment and related ministries and agencies at different levels of government can promote win-win opportunities for environment and development. This integration would not only strengthen the organizational capacity and decision-making influence of environment ministries, but also enhance accountability regarding the potential environmental and social impacts of development projects.

Integration can be supported through impact assessment and monitoring as well as mainstreaming mitigation measures.

Opportunities for integration include decrees, constitutional amendments, presidential committees and independent peer reviews, or environmental focal points within ministries. Some degree of mainstreaming is practised by Australia, China, Japan, New Zealand and the Republic of Korea. Benefits include a more inclusive policy making process, greater coherence of policies and improved implementation. Environmental agendas, however, can be watered down by other stakeholders, and sectoral ministries are not always keen to promote the development projects with the least environmental impacts.

In some cases, achievement of global sustainability goals can be accelerated by allocating authority at more appropriate levels, for example through decentralization and the introduction of participatory approaches in natural resource management. Delegation of authority needs to be linked with appropriate budget authority, human resources, capacity building and reporting mechanisms to ensure effectiveness and public support.

At upstream planning stages, environmental impact assessment for individual projects, cumulative impact assessment for series of projects, and strategic environmental assessment for policies, plans and programmes all provide essential information. Ultimately, fully integrated sustainability assessment is desirable. Benefits include reduced risk of harm to human health or livelihoods due to environmental damage, protection of resources and improved cost-effectiveness of environmental protection measures. Moreover, reduced future remediation costs, political acceptance of the need for preventive and remedial measures, and better institutional collaboration are also benefcial. Assessments have some technical limitations as results are not always well used, and stakeholder participation is sometimes weak; impact assessments are only effective if governments commit to ensuring compliance and enforcement of the changes indicated.

The policy cluster for strengthening incentive structures focuses on appropriate pricing mechanisms. The current fscal incentive structure in many countries does not account for environmental and social externalities and remains a significant barrier to sustainability. Improving the fscal incentive structure can help accelerate greening of the economy to advance the dual goals of economic development and environmental sustainability. Implementation of green taxes can secure a double dividend for the economy and the environment by emphasizing a shift from taxing the "goods" (labour) to taxing the "bads" (unsustainable resource use and pollution). Such a shift would help alter the economic incentive structure in favour of environment-friendly activities, thus reducing trade-offs between economic efciency and maintenance of environmental quality. To be successful, it needs to be implemented as a package and to include awareness campaigns for political acceptability as well as adequate safeguards for vulnerable groups.

The accountability and stakeholder participation cluster addresses the task of integration, which is underpinned by the collection, storage and sharing of information as well as general access to it, such as requiring industries to self-monitor and self-report on environmental and social performance as well as enabling non-governmental organizations to monitor this information independently. Policy options mandating environmental monitoring provide useful information for decision making. Freedom of information laws can enable people to monitor public- and private- sector performance and demand compliance and enforcement action, such as promoted by the Asian Environmental Compliance and Enforcement Network. Access to information promotes public participation in decision making, helping to avoid mistakes and future costs, and increasing political acceptance of remediation measures. Information disclosure policies provide a sound basis for decision making, identifying priorities and formulating cost-effective strategies. Limitations include the difficulties in collecting relevant and reliable data, potential sensitivities in data sharing and the long-term costs of data management.

Policies that require multi-stakeholder participation are also essential to accelerate achievement of the selected global goals. The benefits of multi-stakeholder participation include improved environmental performance, access by disadvantaged groups to redress environmental harm or eliminate potential risks, a better understanding by all parties of the costs and benefits of specifc actions, improved decision making, increased political awareness and active participation, and improved cross-sectoral collaboration. Sometimes institutions accustomed to wielding excessive power resist increased stakeholder participation, necessitating the balancing role of civil society. In the region, the monitoring functions of governments can be usefully supplemented by civil society, especially in situations where government capacity is constrained.

The compliance and enforcement cluster encompasses environmental law and regulation. Imperfect markets, collective action issues and equity concerns characterize many of the environmental issues in the Asia and Pacifc region. A fnal policy cluster, therefore, focuses on access to the environmental judiciary, with policies covering the integrated chain of command-and-control measures, standard setting, monitoring and reporting, investigations and inspection of environmental crimes, establishment of environmental police and "green" courts, and trained environmental judges and prosecutors. Examples include the Environmental Court Act of Bangladesh, the environmental police force in Viet Nam, the green bench of the Supreme and High Courts and the recently established Green Tribunal in India as well as regulatory initiatives to combat corruption in Indonesia and Singapore.

The benefits are to halt or delay potentially environmentally damaging or risky activities, to reach settlements where damage has been caused, to save on future remediation costs if risks are reduced and ultimately to increase social capital. Specialized courts are important as they reduce the burden on enforcement agencies, enable confict resolution and hold government agencies responsible for implementation. Establishment of green courts could derive from popular environmental movements and international treaties. The limitations are that decisions reached through these procedures are not always enforced, and potential jurisdictional conficts may emerge between newly formed enforcement agencies and existing sectoral agencies. Court procedures may also be lengthy and costly, thus effectively excluding disadvantaged groups and poor individuals.

Benefits and Limitations

The main considerations of benefits and limitations included in the policy analysis are as follows:

- economic and financial;
- social;
- environmental;
- political and institutional; and
- distributional.

In general, the priority policies identified here involve up-front costs, which have been a significant barrier in the developing countries of Asia and the Pacifc, where investment priorities have been preferentially directed towards economic growth and poverty reduction rather than environmental goals. The so-called low-hanging fruit, such as energy efciency measures that can be undertaken relatively easily, are still ripe for harvesting in many countries, but may not achieve the scale of structural change needed to contribute to global goals. For example, the transition to a low-carbon society, which has multiple co-benefits, requires massive investment in new, renewable energy systems which are often not the cheapest option under current conditions. In addition, the barriers to adopting energy-efcient technologies can be multi-faceted. The challenge is to change those circumstances, such as through the elimination of fossil fuel subsidies and the accelerated depreciation of existing fossil fuel installations, to create a situation in which renewable energy will be able to compete on equal terms with coal and oil. Continued technological development and cooperation on technology transfer to reduce the costs of this transition are also essential.

The social implications of the priority policies have tended to receive less attention than the economic issues, but a common theme is that improved access to information, stakeholder empowerment, gender equality, access to environmental justice, and equity and benefit sharing will all assist in the transition to sustainable development. Safety nets and compensation for poor households - for example when disadvantaged by removal of subsidies - are also part of the mix. The main barriers are unequal powers and the advantages of elites, conficting social and private values and the attraction and psychological power of conspicuous consumption. Problems of compliance and enforcement are common from the household to the industry level, while higher household costs and employment impacts on declining industries during policy-initiated transitions can hinder progress. The need to relocate vulnerable communities from areas of risk can also present a challenge.

The political and institutional consequences of the suggested priority policies are critical, as it is unlikely that business-as-usual regimes will enable such a policy mix to be effectively implemented. Political benefits are observed from promoting elements of the green economy that promise economic growth, new employment opportunities and environmental benefits. These benefits are best illustrated in the economic stimulus responses to the global financial crisis, where countries such as the Republic of Korea and China have led the world in green stimulus packages. The political downside, however, is reffeed in fears that being a first mover, such as in implementing a carbon tax approach to combat climate change, will lead to unfair advantages for those who lag behind. International coordination, political vision and courage, supported by a willing electorate, are needed to overcome such fears and become a green pioneer.

Institutional changes are commonly observed to be incremental tinkering at the margins rather than wholehearted reform. For example, many countries in Asia and the Pacifc have added climate change responsibilities to an existing environmental agency rather than seeking widespread institutional reform that would mainstream climate change across all existing agencies. Sub-regional agencies such as the ASEAN Centre for Biodiversity and the ASEAN Centre of Energy have been created, but without secured funding to support implementation of their mandates. Overall, it appears that institutional change in Asia and the Pacifc has been *ad hoc* rather than systematic.

The distributional impacts of priority policies that determine who wins and who loses have been barely investigated in Asia and the Pacifc. In the energy sector, fossil fuel industries stand to lose market share, although leading companies may ultimately make the transition to alternative energy. In relation to biodiversity, new funding arrangements such as payment for ecosystem services including REDD+ suggest that traditional forest-dependent communities should benefit, but there is potential for powerful groups to manipulate such schemes and for traditional users to become losers. In the freshwater sector, agriculture tends to lose out to urban and industrial water users, as they are prepared to pay higher prices. In waste recycling, traditional waste collectors and the informal sector are beginning to lose out to more formalized waste management, as the value of material resources embedded in waste rises in response to shortages of raw materials.

Transfer and Replication of Successful Policies

As most successful policies have had demonstrable success in a limited range of countries, some only in developed countries, others in large economies, the objective is to analyse whether the suggested priority policies are readily transferable and replicable across national boundaries. If they cannot easily be transferred because they depend on local circumstances that are not widely prevalent, they are unlikely to accelerate the achievement of the selected global goals. The factors considered include:

- how many countries have already implemented such policies;
- how quickly the policies have been adopted by multiple countries since their first introduction;
- how easily the private sector has been convinced that the policies are not harmful to their businesses; and

- how the policies have contributed co-benefits that made them even more acceptable.

Internal Replication

Policy responses in the Asia and Pacifc region increasingly need to adopt a cross-sectoral approach. Even though the identification of the recommended priority policies in one sector helps focus attention, policy interventions are needed for simultaneously addressing several of the environmental challenges listed, and packages of complementary policies should combine to achieve desirable environmental outcomes across sectors. For instance, environmental governance for climate change should not be seen as separate from that for water, and environmental policies for water cannot be separated from agriculture and food security because there are interactions between those concerns as well as with climate change. Policies that build a community's capacity to adapt to or deal with water insecurity in general will facilitate their adaptation to the effes of climate change. For example, construction of small reservoirs in upper watershed areas can provide both increased water security for downstream villages and drought and food control to combat climate change. Climate change never has effes in isolation from those of other drivers or pressures of environmental change, but tends to aggravate the effes of existing ones. What seems like a priority policy intervention today might, over the long run, turn out to be maladapted from a climate change point of view, thus necessitating constant learning from experience and flexible, adaptive management.

External Replication

Many of the policies being adopted in the Asia and Pacifc region had their origin and initial trials in other regions, often Europe and the United States. The failure to implement many policies in this region successfully may stem from over-optimism that if a policy already works in a developed country then it should also work in a developing one. For example, the strong command-and-control policy regime to manage air and water pollution in the United States, involving standard setting, permits and prosecution of offenders, tends not to work as well in the developing countries of Asia and the Pacifc. A policy regime built around voluntary compliance, the social pressures of naming and shaming polluters, and compensation of victims may be more applicable for the socio-cultural context of the region, although measures of effectiveness require further analysis.

Creating an Enabling Environment for Effective Policies

Part of the above analysis relates to the enabling and/or impeding factors that have led to the success or otherwise of specifc policies. It starts from the premise that policies are only as good as their implementation. Theoretically brilliant policies that are complex or difficult to implement are unlikely to be easily transferable or replicable across national boundaries.

Moving from Successful Policies to Policies Plus Enabling Conditions

Identifying policies that work, and those that do not, is of little relevance unless the underpinning causal conditions and context are understood. A growing body of policy studies identifes the factors that allow policy change to happen. There are many critiques of why policy change in the irrigation sector in South Asia, for instance, has been slow and tardy. Accordingly, it is useful to look at what policies have failed to accomplish and what they have achieved.

Filling the Policy Vacuum

Finally, it is important to look at the changing configuration of actors like the state, markets and civil society. Increasingly, there is a policy vacuum left by the state that the market and/or civil society may intervene to fll. For instance, a common response to the inadequate supply of water in many cities in Asia and the Pacifc is the rising popularity of water markets – the sale of safe water by private entrepreneurs – seen in cities like Chennai and Kathmandu, amongst others. Privatization of water markets can increase efciency, but laws should be put in place to anticipate and limit monopolies.

Given the challenges of balancing growth, poverty alleviation, environmental protection and resource conservation, governance improvements should prioritize addressing environmental concerns at the macro-economic level, specifically focusing on patterns of growth and restructuring the economy. Environment remains the foundation of sustainable development and if it is weak or compromised, then the social and economic pillars of sustainable development will also be undermined. By allowing co-benefits, ancillary costs and environmental externalities to be integrated into social and economic decision making, environmental concerns and climate change can be mainstreamed into a sustainable development framework. Carefully monitoring the actual outcomes of policy decisions and making continuous adjustments, with the full input of all stakeholders, are key elements of an adaptive management approach. The recent efforts to promote the green economy and green growth in the region suggest that the time is ripe to begin this transformation.

The implications of failing to implement the recommended policies or clusters of policies make a strong case for their adoption, as improved environmental governance is necessary to reverse environmental degradation and the unsustainable use of natural resources. There is adequate experience with several of the priority policies analysed above to justify faster replication, but the difering governance contexts and enabling environments in a region as diverse as Asia and the Pacifc may be barriers to adoption. Each of the suggested priority policies or policy clusters has experienced some limitations during implementation, so these challenges also need to be better understood, and additional basic research needs to be undertaken before specifc recommendations can be made on the transfer of policies to new jurisdictions.

References

ACB (2011). *Legal and Institutional Development for Promoting Access and Benefit Sharing of Genetic Resources in Southeast Asian Countries*. ASEAN Centre for Biodiversity, Los Banos

IGES (2008). *Climate Change Policies in Asia-Pacific: Re-Uniting Climate Change and Sustainable Development.* White Paper. Institute for Global Environmental Strategies,

Medina, M. (2007). *The World's Scavengers: Salvaging for Sustainable Consumption and Production.* Alta Mira Press, Lanham, MD

OECD (2006). *Good Practices in the National Sustainable Development Strategies of OECD Countries.* Organisation for Economic Co-operation and Development, Paris.

Shekdar, A. (2009). Sustainable solid waste management: an integrated approach for Asian countries. *Waste Management* 29(4), 14381448

World Bank (2006). *Where is the Wealth of Nations? Measuring Capital for the 21st Century*. World Bank, Washington, DC

10

Environmental Governance Mechanisms in Europe

The pan-European region is very diverse, with its 37 different national languages spoken in the 50 European countries and its range of socio-economic and political systems, as well as in its varied physical environment and means of environmental governance. Europe's land area of 23 million km^2 is characterized by a great variety of (agri)cultural landscapes, urban agglomerations, extensive coastal zones, forests and undisturbed pristine areas. Of the nearly 833 million Europeans, about half live in Western Europe, while some 72 per cent of the entire region's population resides in urban areas.

Conversion to and intensification of agriculture along with increasing demand for greater mobility and urban space have transformed a majority of European landscapes over the past 100 years, causing fragmentation and loss of natural and semi-natural habitats with an associated decline in biodiversity.

Indeed, considerable progress has been made in meeting environmental targets, with the situation improving in many areas. Nonetheless, concerns about long-term threats to the environment and human health persist, the latter especially for Europe's large urban population. Despite some successes in decoupling environmental pressures from economic growth, Europe's environmental footprint remains disproportionately high. This is due to the continued unsustainable use of natural resources both within and outside the region to satisfy the high production and consumption level of its inhabitants.

These trends are increasingly linked and complex, and require an integrated policy approach for which strong governance mechanisms need to be in place. Given that Central and Western Europe in particular have a dense network of political boundaries, a regional focus to tackle environmental issues is necessary. A major attribute of the pan-European region is its economic and political interconnectedness, combined with well-established and strong formal governance mechanisms and structures to address environmental issues at (sub-)regional level. This has made Europe a leader in transboundary as well as global environmental decision-making. In particular, the EU has more than four decades of experience in developing environmental policies: the first Environmental Action Programme (EAP) was adopted in 1972 while the sixth ends in mid-2012. EU legislation is implemented at a national level within EU Member States, with forceful implementation control by EU institutions. The legislation is also used in non-member European states on a case-by-case and voluntary basis.

Policy approaches have evolved over time, using policies and single-issue instruments in the 1970s and 1980s, followed by policy integration and raising public awareness in the 1980s and 1990s and thereafter. An integral part of EU environmental governance is the regular monitoring, reporting and assessment required by EU legislation; these activities help to inform policy makers about effectiveness and also help to identify emerging issues. Since the early 2000s, European environmental policy has increasingly been guided by the fact that well-designed, coherent policies that integrate different sectoral policy domains can provide greater benefits at lower costs than several single policies. As a result, Europe's natural resources are used with increasing efciency.

This concept is already being emulated in EU neighbouring countries and in the pan-European Environment for Europe ministerial process that was initiated in 1991. In September 2011, for example, the Seventh pan-European Environment for Europe Ministerial Conference focused on the sustainable management of water and water-related ecosystems and on greening the economy including mainstreaming the environment into economic development.

The countries of Eastern Europe also have well-developed formal environmental policies and regulations, although the implementation and enforcement of these has often tended to lag. In the early 1990s, following the collapse of industry in Eastern Europe, environmental pressures dropped substantially in many countries, giving the public and authorities a false sense of security. The focus of attention shifted towards more urgent needs related to economic restructuring and development, with an inclination to make the economic transition easier by reducing environmental regulation. At first this strategy worked, but later, when countries regained their economic strength, it began to backfre.

A new wave of improved environmental legislation and policies can now be expected in the non-EU part of Europe, the current global financial crisis notwithstanding. Promising policies include, for example, integrated river basin management and cross-boundary biodiversity conservation. Another example is the Inter-Parliamentary Assembly of the Commonwealth of Independent States (IPA CIS), with its consultative and informative role. It has a Permanent Commission on Agricultural Policy, Natural Resources and Ecology, which advises the parliaments of CIS countries and suggests sample legislation on environmental issues. Practically all aspects of modern environmental policy are covered, ranging from environmental security, environmental insurance and strategic environmental assessments to environmental monitoring, energy conservation and environmental education.

Policy Appraisal

The five key challenges/priority issues were identified for Europe, in no particular order, during a GEO regional consultation held in September 2010:

- climate change;
- air quality;
- freshwater;
- chemicals and waste; and
- biodiversity.

At the GEO Regional Consultation, five international environmental goals related to the key challenges were identified, and regional-level goals were added later where applicable. The group then selected promising policies that have already shown some degree of success in helping to speed up meeting the globally and regionally agreed environmental goals.

The Introduction to *GEO-5* explains the methodology applied in this appraisal in more detail. It is acknowledged that:

- recent, innovative policies do not have a long enough track record to be selected for this appraisal, though some promising emerging policies are included in the conclusions at the end it;
- even where actual evidence of policy effectiveness does exist, such environmental improvements cannot usually be directly linked to one single policy or cluster of policies because of impacts from other sectoral policies, economic developments or political restructuring; and
- there are certainly other priority issues in parts of the region - such as the marine and coastal areas surrounding large parts of Europe, a new energy mix, land-use change and land degradation, or developments in Europe's mountainous areas - but these were not among the maximum of five key challenges/priority issues selected for this analysis by the GEO regional consultation.

Climate Change

In terms of the total reduction in greenhouse gas emissions, European countries are leading the global climate change mitigation effort by a wide margin. Other large advanced economies have either not ratified the Kyoto Protocol (United States), are failing to meet their Kyoto targets (Canada) or are allowed to increase their emissions (Australia). Japan, with its 6 per cent reduction target, is the main exception.

The EU-15 is well on track to meet its Kyoto target; indeed, over-compliance may even be achieved when the Clean Development Mechanism, Joint Implementation mechanism, and carbon removals such as forestry activities, are factored in. None of the Central and Eastern European countries have faced any problems in meeting their Kyoto Protocol obligations as their targets were set before the fall in emissions associated with the collapse of the Soviet bloc. In addition, regional emission targets for the post-2012 period have been set.

Large-scale reductions of greenhouse gas emissions can only be achieved, however, through a tightly coordinated and coherent combination of different policies targeting different economic sectors and sources of emissions. Only then can efcient synergies be achieved.

In 2009, the EU formally adopted its climate and energy package, an integrated approach with binding legislation to implement the EU's three main climate and energy targets:

- reduce EU greenhouse gas emissions to 20 per cent below 1990 levels by 2020;
- reduce EU energy consumption by 20 per cent compared with projected levels by 2020 through increased energy efciency; and
- meet 20 per cent of EU energy needs from renewable sources, including biofuels, by 2020.

These commitments together have been labelled the 20-20-20 targets, which are being implemented through an array of policies ranging from carbon taxes and emissions trading schemes to local voluntary efforts by municipalities. Two of the most promising policies are discussed below.

European Emissions Trading System

The EU Emissions Trading System (EU ETS) was launched in 2005 as a cornerstone of EU climate policy and the key tool for reducing industrial greenhouse gas emissions in a cost-effective manner. It is the first and largest international scheme for trading emission allowances, and is open to non-EU countries on the condition that they meet the strict standards of the EU ETS.

The EU ETS covers about 40 per cent of EU greenhouse gas emissions. For 2009, the EU carbon trading market was estimated to be worth nearly US$118.5 billion per year, compared to a global carbon credit market worth an estimated US$143.75 billion. In 2009, the volume of emissions covered by the system reached 6.33 billion tonnes compared to 41 million tonnes covered by the Chicago Climate Exchange and second (2008–2013) phases of the EU ETS, although caused by different factors, demonstrate the crucial requirements of the supply of accurate, reliable and constantly updated figures on energy consumption and emissions, verified by strict monitoring. The Third Trading Period will implement several important changes such as inclusion of airline emissions, increased auctioning of allowances, and an ambitious EU-wide cap instead of national caps. The EU ETS cap will decrease continuously from 2013 onwards using a linear reduction factor.

The EU ETS is an attractive option for European countries outside the EU. Iceland, Liechtenstein and Norway are already covered by it through their membership of the European Economic Area Agreement, while Switzerland will be the first non-EU country whose national emissions trading system is linked to the EU ETS and Australia is exploring this possibility as well.

Feed-in Tarifs for Renewable Energy Systems

Feed-in tarif schemes were elaborated as the main support mechanism for renewable energy systems. Its goal goes far beyond reducing carbon dioxide (CO_2) emissions only; it also takes issues such as energy security, independence from conventional fuel price volatility or decentralization of energy into account.

The policy ofers long-term contracts to renewable energy producers, typically based on the cost of generation of each technology with two basic pricing models: a market-independent fxed price, applied by Germany's Renewable Energy Source Act, and a market-dependent premium price model, used, for example, by Spain. The German Renewable Energy Feed-in-Tarif (REFIT) scheme, launched as early as 1991, is a successful example. Spain is another positive example, as the country has established a dynamic, export-oriented and job-creating renewable energy sector even if it has not succeeded in other areas of climate policy. About two-thirds of EU Member States have now built up renewable energy capacity using feed-in tarifs.

At least 17 developing countries and emerging economies, including Brazil, China, India, Kenya, Nicaragua, South Africa and the Republic of Tanzania, have feed-in tarif schemes in place, most of which have been implemented within the last five years through, among others, the Global Energy Transfer Feed-in Tarif for Developing Countries (GET FIT) programme. About 60 per cent of projects that have been registered under the Clean Development Mechanism or are in the pipeline for 2012 are for renewable energy, showing that development of this has become the most attractive climate policy option for developing countries.

Climate Adaptation Policies

When foods caused serious human and material damage in Central Europe in the summer of 2002, the European Commission (EC) reacted immediately by proposing the use of existing funds in a flexible way to respond to the urgent needs of the people affeed. By mid-November 2002, an EU Solidarity Fund (EUSF) had been launched to fnance short-term responses such as reconstruction of damaged or destroyed infrastructure, and to secure protective infrastructure such as dams and dykes. As the EUSF is restricted to the uninsured sectors of public infrastructure, it should be supplemented with a unified, innovative insurance system, being developed across Europe, which has the power to transfer risks from the local level to national and even global insurance markets through primary insurance and re-insurance.

Another tool to assist in preparing for the impacts of climate change is the 2007 EU Floods Directive, under which draft national food risk maps had to be submitted in 2011, with fnal versions to be ready by 2013 and fnal adaptation plans by 2015. More recently, the EU White Paper on Adaptation to Climate Change has moved beyond short-term disaster responses, outlining key steps towards a European framework for long-term adaptation measures and policies to increase resilience, to be implemented at national and local levels. Top-down strategies are envisaged for mainstreaming adaptation into sectoral policies, focusing on sectors such as land-use planning, agriculture, water management and biodiversity/nature conservation. Bottom-up activities focus on building adaptive capacity and implementing action at municipal level. In addition, a new EU Clearinghouse on climate change impacts, vulnerability and adaptation was put in place with the first stage of the strategy to run until 2012.

Air Quality

Although many aspects of air quality across Europe have improved in recent decades due to emission reductions from industry and transport, air pollution continues to pose a threat to human health, especially in urban areas. For example, exposure to fne particulate matter ($PM_{2.5}$)was estimated to have caused 5 million lost life years in 2005in the EEA-32. Similarly, other air pollutants continue to cause environmental damage to ecosystems, with 10 per cent of the EEA-32 natural ecosystem area still subject to acidifying pollutant deposition caused by sulphur dioxide (SO_2) and nitrogen oxides (NO_x), and more than 40 per cent of sensitive terrestrial and freshwater ecosystems still subject to eutrophying atmospheric nitrogen deposition in the form of nitrogen oxides and ammonia (NH_3). Despite declining peak ground-level ozone (O_3) concentrations, background levels are steadily rising, also leading to ecosystem damage.

The Convention on Long Range Transboundary Air Pollution (CLRTAP) of the United Nations Economic Commission for Europe

(UNECE) has been pivotal in providing the scientifc evidence that has underpinned efforts to shape air quality policy. CLRTAP's fagship 1999 Gothenburg Protocol promotes an integrated multi-pollutant, multi-effe approach to optimize efforts to improve air quality across Europe. It is comparable to the 2001 EU National Emission Ceilings Directive, which establishes legally binding pollutant-specifc emission ceilings for nitrogen oxides, non-methane volatile organic compounds, sulphur dioxide and ammonia for the EU-27. The 2008 Clean Air for Europe (CAFE) Directive merges much of the existing air quality legislation to develop long-term, strategic and integrated policy advice.

Such European approaches have been instrumental in providing the impetus for the development of a suite of air quality policies through the establishment of binding emission and air quality standards. Three outstanding environmental success stories are described here: vehicle emissions and fuel standards, the EU Industrial Emissions Directive and local air quality management policies.

European Vehicle Emission and Fuel Standards

Historically, road transport has contributed substantially to atmospheric pollution by producing emissions of lead (Pb), nitrogen oxides and particulate matter. Reduction of these emissions has been achieved through the establishment of EU directives controlling both fuel and vehicle emissions, with fuel policy focused on banning lead and limiting sulphur content. European Vehicle Emission Standards (Euro standards) control exhaust emissions of nitrogen oxides, non-methane volatile organic compounds and total hydrocarbons, carbon monoxide and particulate matter from new vehicles sold within the EU. Since the establishment of the Euro 1 standards in 1992, more stringent ones have been introduced, tightening controls on different pollutants, vehicle categories, weights and classes, engine volumes and fuel types; Euro 5 standards have been in force since 2007.

Despite a 26 per cent increase in fuel consumption across the entire transport sector between 1990 and 2005, actual pollutant emissions in 2005 were significantly lower than a theoretical no-new-policy scenario assuming conventional technologies and no introduction of Euro standards: in the EEA-32, nitrogen oxide was 40 per cent lower than the scenario figures; carbon monoxide 80 per cent, non-methane volatile organic compounds 68 per cent and particulate matter 60 per cent lower. Lead emissions from road transport alone decreased by 99 per cent and sulphur dioxide emissions by 92 per cent between 1990 and 2008. Additional benefits of the Euro standards include increased engine lifetime and lower maintenance costs due to the removal of sulphur, better fuel economy and reduced greenhouse gas emissions.

The implementation of cleaner fuel policies involves costs related to fuel switching, such as replacing lead with other fuel additives and shifting to producers of low-sulphur crude oil, improving engine technologies and upgrading refneries. However, the benefits of lead phase-out and desulphurization in terms of human and ecosystem health in general exceed the costs. The EU, Japan and the United States lead the world in desulphurization policies and by 2011, vehicle fuel in the European region was lead free.

The time lag in the effectiveness of vehicle emission policy depends on the average age of the vehicle feet and the afordability of new vehicles. Awareness raising, product labelling, enforcement and regular control of fuel quality, now considered essential for vehicle policies to reach their full potential, have ensured successful implementation of these policies.

EU Industrial Emissions Directive

The 2010 EU Industrial Emissions Directive is designed to consolidate seven existing EU directives that have evolved since the early 1980s and have been instrumental in reducing industrial sulphur dioxide emission. The new directive will combine proven policy measures including technical emission controls, best available techniques, fuel switching and reduced sulphur content in liquid fuels.

Implementation of these measures has resulted in a clear reduction in sulphur dioxide emissions across Europe over recent decades, effectively decoupling them from industrial activities, especially in Western Europe. To some extent these reductions were aided by socio-political and economic changes between 1990 and 2000 in former socialist Eastern Europe. The reduction of total anthropogenic sulphur dioxide emissions in the EU-27 – by 80 per cent between 1990 and 2009 – has led to substantial declines in acidification rates as exceedances of critical loads across Europe have been reduced. The implementation of the measures did, however, involve additional costs, requiring investment from the private and public sectors.

The new EU Industrial Emissions Directive aims to reduce these costs by streamlining and enhancing cost efciency and effectiveness. Many control technologies have proven transferability, having been adopted in many Asian countries where they are of particular relevance since 80 per cent of Asia's energy demand is met by coal-fred power. Enhanced penetration of measures across Asia could yield further substantial improvements in sulphur dioxide emission reductions.

Local Air Quality Management Policies

Under the 2008 Clean Air for Europe (CAFE) Directive, local authorities are obliged to prepare air quality management plans to ensure compliance with air quality standards. Many policy measures have focused on urban transport since this sector generates 70 per cent of air pollutants in urban areas. Perhaps the most influential of these policy measures have been low-emission zones that limit or ban the most polluting vehicles from entering urban areas and encourage faster renewal of the vehicle feet in line with vehicle emission standards. Around 100 low-emission zones in ten European countries have either been established or are in the process of being established. Other measures include congestion charging, expansion and improvement of public transport and cycling infrastructure, car pooling and cycle sharing systems, renewal or retroftting of municipal vehicle feets, and trafc and green areas management. Air quality management plans also require the public dissemination of current information detailing ambient air pollution and exceedances of air quality standards, with citizens and legal entities having the right to go to court in cases of non-compliance with standards. However, individual lawsuits for breaches of air quality standards are rarely pursued due to costs, time-consuming procedures and low awareness. Besides, many urban areas in Europe are not compliant with current European air quality legislation. To be really successful, local air quality management plans require adequate monitoring and information systems and appropriate institutional mandates for local authorities.

Freshwater Management

In large parts of Europe demand for water often exceeds local availability, a trend that is likely to be exacerbated by climate change. In addition, both point and diffuse sources of pollution are still significant in parts of Europe, as a result of which some health risks remain. Europe's water challenges are driven by competing demands for water by agriculture, industry, public water supply and tourism, and further complicated by the transboundary nature of many European freshwater resources. Addressing these challenges calls for strong environmental governance structures, with a focus on coherent and integrated efforts and regional cooperation.

The EU Water Framework Directive and pan-European UNECE instruments such as the Convention on the Protection and Use of Transboundary Watercourses (Water Convention) provide the basis for solving major water issues in the region. The Water Framework Directive brings the many isolated policies that have been developed in the EU since the mid-1970s together into one coherent legal framework for water policy decision-making within the river basin context. Its main goal is to protect and enhance the status of all EU waters, including groundwater, rivers, lakes and coastal waters, and water-dependent ecosystems, and to ensure long-term sustainable use of water resources. The Water Convention ofers a common platform for both EU and non-EU European countries to exchange and transfer knowledge and create common understanding, and is a useful tool for assisting in implementing EU water legislation by non-EU countries.

Three specifc policy instruments with some history of effective implementation have been selected for further appraisal: integrated management of transboundary river basins; policy mixes to address non-point sources of pollution; and water metering and volume-based pricing.

Integrated Management of Transboundary River Basins

Water does not stop at administrative or political boundaries, making regional cooperation crucial between countries that share the natural geographical and hydrological unit of a river basin. The overarching approach of integrated water resources management has proven to be an effective policy for valuing, managing and protecting water-related ecosystems. The development of river basin management plans is one of the main tools for implementing the Water Framework Directive, which focuses on pollution prevention and control, greater public participation in water management, and economic analysis of water use. The plans require the integration of industrial, agricultural and rural development, nature conservation and forestry programmes at the river basin scale and, in many cases, transboundary collaboration and coordination through river basin commissions. Progress in cooperation in the region is varied, however.

The first river basin commission in Europe - the International Commission for the Protection of the Rhine - celebrated its 60th anniversary in 2010, and has registered numerous successes over the years. Similar commissions have been established since for many European rivers, gradually moving eastwards despite the fact that in non-EU countries a comprehensive and strong legal basis for cooperation is often still lacking. As many water bodies are shared by EU and non-EU nations, countries are encouraged to jointly prepare river basin management plans: the Tisza River Basin Management Plan provides a recent example of such cooperation across EU borders.

By sharing the benefits and responsibility of sustainably co-managed water resources, economic development is fostered, establishing a connection between economic activities and the environment. River basin management plans also encourage public participation in working and expert groups. However, this approach still faces serious limitations due to the magnitude and complexity of the problems it seeks to address and the significant number of stakeholders who need to be involved.

Policies to Address Diffuse Sources of Water Pollution

Eutrophication, predominantly caused by sewage discharges and agricultural run-of, is a major threat to European freshwater resources. Policies to reduce the fow of nutrients from point sources are well known and have proved to be successful, provided that sufcient funding is allocated to construct and manage water treatment systems. Tackling the problem of diffuse sources of freshwater pollution is much more challenging. There is significant experience in Europe of applying mixes of policies to reduce diffuse nutrient run-of, including accounting systems for the use of nitrogen in agriculture, regulations on livestock density and the use of animal manure, purchase of nitrogen quotas, taxes on fertilizers, and compensation for converting agricultural land into wetlands or forest. Denmark, for example, has applied a large cluster of such mixed policies since the late 1980s, taking their synergistic effes into account, while avoiding disproportionate burdens on any particular stakeholder. As a result, the application of nutrients in Denmark has steadily decreased since the early 1990s.

Water Metering and Volume-based Pricing

Europe has relatively high water consumption due to high agricultural and industrial demand. In addition, considerable losses often occur in the supply chain, aggravating shortages in already water-scarce regions. In some countries up to 40 per cent of the total amount of water transported may be lost before it even reaches the consumer, while in others it can be below 10 per cent. Metering, cost recovery tarifs and proper pricing structures stimulate more responsible water use, at the same time generating funds for the maintenance of the supply system.

While water metering is a common policy tool in many counties of Western Europe, Central and Eastern European countries are still in the process of transition from a fat-rate price charged per person to a system of volume metering. Various studies reveal that on average, if individual metering systems are in place, reductions of 10–40 per cent can be achieved in household water use.

In addition to metering, several Western European countries apply cost recovery tarifs and have introduced site-specifc pricing structures. An increasing block rate creates the strongest incentive for conservation, applying a user-pays principle, under which the unit rate for water increases with water use, keeping the price for basic water needs relatively low. This system is becoming more common in both households and commercial sectors in Western European countries.

Applying this experience in Central and Eastern Europe would not only reduce inefcient water consumption, but also generate funds for modernizing the water sector, increasing the reliability of water services delivery.

This policy, however, faces several limitations. The costs of meter installation could be too heavy a burden for poor households, conficting with the Millennium Development Goal (MDG) 7c of halving, by 2015, the proportion of the population without sustainable access to safe drinking water and basic sanitation.

Also, water pricing should not result in a situation in which personal hygiene and health are compromised in order to pay a water bill. To become successful, water pricing and meter installation require a good understanding of relationships between price and use in each sector, taking local conditions into account. Special subsidy schemes could be introduced for providing free meter installation for poor families, gradual repayment terms, and special provisions for writing of vulnerable families' accumulated water debts.

Preventions of Chemicals and Waste

Both in the EU and in Eastern Europe, issues related to chemicals and waste have always been of primary importance. The EU's waste policy consists of three levels of legislation. The first, so-called horizontal, level defnes overall requirements for all waste types and consists of the 2008 Waste Framework Directive, which is the cornerstone of current EU waste policy, and the 2006 Waste Shipment Regulation. The second level of legislation deals with waste installations and includes the Waste Incineration Directive, the 1999 Landfll Directive and the 2000 Port Facilities Directive. In addition, the 2010 Industrial Emissions Directive also defnes requirements for some waste installations. Finally, the third level deals with specifc waste streams such as waste containing polychlorinated biphenyls and terphenyls (PCBs/PCTs), waste oils, sewage sludge, electrical accumulators like batteries, and packaging waste. One example of such regulation is the Waste Electrical and Electronic Equipment Directive, which deals with the collection and recycling of such waste. It also includes the Restriction of Hazardous Substances Directive, which bans the use of certain hazardous substances in electrical and electronic products.

Ukraine all have large amounts of industrial waste in landflls as well as mining waste, with few or no financial incentives to recycle them. This is the result of many Soviet-era waste management and reuse practices being abandoned without alternative schemes being introduced.

Waste Prevention

The EU Waste Prevention Directive of 2008 is based on defnitions laid down in the Waste Framework Directive, in which prevention has been given the highest priority. Article 3.12 of the directive demands waste prevention through measures taken before a substance, material or product has become waste, by reducing:

- the quantity of waste, including through the reuse of products or the extension of the lifespan of products;
- the adverse impacts of the generated waste on the environment and human health; or
- the content of harmful substances in materials and products.

Waste prevention should also incorporate such aspects as ecodesign, life-cycle approaches, changing business models and consumption patterns.

One of the EU's major aims has always been waste reduction, but this goal has so far not been achieved. On the contrary, the amount of waste has been growing; notable examples including construction and demolition waste, packaging, hazardous and municipal waste, and sewage sludge. This trend has to be reversed, particularly as resource efciency is one of seven fagship initiatives of the EC's Europe 2020 Strategy, which is reffeed in the EU goals of decoupling resource use from economic growth, measured as lower resource use per unit of gross domestic product (GDP), and of minimizing waste. In addition to reducing waste generation, it is important to improve waste recycling. Current EU data indicate that only 38 per cent of total waste is reused or recycled.

Even though radioactive waste is not a subject of the waste hierarchy, it has important implications for both safety and energy production. On 19 July 2011, the European Council adopted the Radioactive Waste and Spent Fuel Management Directive, which sets standards for the safe disposal of spent fuel and radioactive waste from nuclear power plants as well as from medicine or research. This was a major achievement achievement for nuclear and environmental safety in the EU.

Other non-EU European countries also face significant challenges in waste policy. For example, Belarus, the Russian Federation and The practical outcomes of this policy can be achieved through a number of instruments, including legal provisions, voluntary agreements, economic instruments and incentives, and communication strategies.

Reuse and Recycling

The Waste Framework Directive also encourages reuse, recycling and recovery, providing a range of options for the recycling of various materials including promoting the establishment of recycling targets, which can be material-specifc. The figures show that the average amount of waste per citizen in the EU is approximately 6 tonnes per year. Municipal solid waste alone increased from 468 kg per person in 1995 to 524 kg in 2008, an increase of 12 per cent, caused by the increasing adoption of Western consumption habits in the new Member States. However, EU countries have made measurable progress in the efcient use of resources and management of waste, as illustrated by the fact that municipal waste recycling more than doubled between 1995 and 2008, rising from 17 per cent to 40 per cent.

Despite such advances, the EU is still not a recycling society, given that as of 2008, the share of municipal waste disposed of in or on land still exceeded 40 per cent. Based on various macro-economic scenarios, it is estimated that by 2035 total waste generation in the EU-27 will have increased by 60–84 per cent compared to 2003 levels, although these figures could be significantly revised due to the current economic crisis.

Eastern Europe shows quite a different picture. In the Russian Federation with a population of nearly 143 million, the total amount of waste generated annually is larger than that of the entire EU with a population of 502 million (3.4 billion tonnes and 2.6 billion tonnes respectively) with 90 per cent of waste originating from the mining industry. On average, however, only about 26 per cent of waste is recycled. Of this recycled waste, 35 per cent is accounted for by industrial waste and only 4-5 per cent by domestic waste. All the other types of waste are effectively not recycled at all.

A life-cycle approach to waste management could significantly reduce Europe's dependence on imports of raw materials and energy consumption for manufacturing new materials. More significant gains can be made, but only through full implementation of the EU's waste directives and in particular the EU Landfll Directive. Reuse and recycling would also require significant changes in consumer behaviour, which could be helped by information and education campaigns.

Chemicals Policy

The most profound and ambitious piece of legislation regulating chemicals in Europe entered into force on 1 June 2007. This legislation deals with the Registration, Evaluation, Authorisation and Restriction of Chemical substances (REACH), and replaces a patchwork of previous directives and regulations. The seven objectives that are essential for achieving a sustainable REACH framework are:

- the protection of human health and the environment;
- the maintenance and enhancement of the competitiveness of the EU chemical industry;
- the prevention of fragmentation of the EU's internal market;
- increased transparency;
- the integration with international efforts to regulate the use of chemicals;

- the promotion of non-animal testing; and
- conformity with EU international obligations under the World Trade Organization (WTO).

One of the most important elements of REACH is the registration of chemicals. REACH requires companies that make and/or import chemicals to submit registration dossiers to the European Chemicals Agency (ECHA). The 2010 registration deadline was related to bulk chemicals supplied in quantities of more than 1 000 tonnes per year and very hazardous chemicals; by the REACH deadline of 30 November 2010, the agency had received 24 675 registration dossiers for 4 300 substances. Despite significant concerns raised by the chemicals industry about the unprecedented burden REACH placed on companies and some initial technical difficulties, the overall registration process was a success. The future deadlines in 2013 and 2018 cover chemicals supplied in smaller quantities. Additionally, REACH includes some limited provisions for the integrated assessment of cumulative risks from multiple substances and other stressors.

It is expected that implementation and compliance with this legislation will lead to more predictable markets and a reduction in companies' liabilities, especially by providing a level playing feld for all market players.

Among new developments is the new EU Toy Safety Directive (2009/ 48/EC), with Member States expected to have had the new measures under way as of July 2011, and further parts of the directive coming into force in July 2013. Toys come under REACH regulations, and the new safety directive focuses in particular on limiting the amounts of certain chemicals that may be contained in materials used for them. Additionally, in 2013, the EU will implement the new regulation on chemical substances in cosmetics (1223/2009/EF), aimed at simplifying procedures and streamlining terminology. It will also include new provisions for nanomaterials and endocrine-disrupting substances.

The limitations of all these policy options are partly related to difficulties in obtaining information on the environmental and health risks of chemicals, especially new ones for which the risks are unknown. As there may be business issues related to the cost of flling knowledge gaps and clarifying uncertainties, there could be substantial additional advantages in sharing information between the European Chemicals Agency and its counterparts in transitional European and developing countries.

BIODIVERSITY CONSERVATION

Europeans are at the forefront of establishing multi-national conservation efforts. A wealth of biodiversity conservation policies and tools, including various regional conventions, have been applied to European terrestrial and marine ecosystems. At supra-national level, biodiversity conservation is mainly driven by such EU legal instruments as the Nature Directives adopted in 1979 and 1992 and the pan-European Biological and Landscape Diversity Strategy adopted at the Third Ministerial Environment for Europe Conference in 1995. Although the EU directives are legally-binding and Pan-European strategy is not, the two are mutually supportive and lead to an improved state of biodiversity in Europe. In 2001, the EU and its Member States committed to halt the loss of biodiversity by 2010, but this target was not met, and the status of biodiversity is still a cause for serious concern. As a result, a new EU 2020 biodiversity strategy was endorsed in May 2011.

For the purpose of this analysis, three policy clusters were identified as being benefcial in achieving biodiversity conservation goals:

- the establishment of ecological networks as a key means of reducing biodiversity loss;
- payment for ecosystem services as an instrument for conserving European agro-biodiversity; and
- the sustainable management of forest resources.

Three cases were selected for further appraisal: the EU Natura 2000; agro-environment measures; and the voluntary pan-European Forest Europe process.

THE NATURA 2000 NETWORK

Natura 2000, a tool used by the EU 2020 Biodiversity Strategy, represents the largest supra-national network of protected areas in the world. It incorporates sites established under the EU Habitats and Birds Directives and aims to assure the long-term survival of Europe's threatened and most valuable species and habitats. It has developed steadily over the last 15 years, and is now made up of more than 26 000 sites covering 18 per cent of the EU's land and sea areas. Similar network approaches also apply beyond EU borders.

The Natura 2000 network helps protect vulnerable habitats and species as well as a wide range of ecosystem services, including the regulation of climate (such as mitigation of climate change), purification of water and maintenance of water fows, preservation of landscape and amenity values, and support of tourism and recreation. Furthermore, it facilitates cooperation beyond national boundaries, contributes to the decentralization of national-level conservation policies, and encourages local and regional economic development by ofering job opportunities and helping to attract fnance. Even though implementation of the network requires around US$8 billion (•6 billion) annually, there are several examples demonstrating that the benefits exceed the associated costs.

While the development of the network has made little headway with marine environments, it is a real success for terrestrial ecosystems. The conservation status is, however, still only favourable for less than 20 per cent of terrestrial habitats and species, both within and outside the Natura 2000 network. Initially, the designation of sites faced a number of problems, but these are being overcome through the democratization of multilevel biodiversity governance. To avoid many sensitive problems in negotiations, for example, in 1997 the EC initiated an apolitical process for selecting sites in a bio-geographical context through scientifc seminars where boundaries were agreed.

Agri-environment Measures

The need to preserve high nature-value farmland in the EU was agreed in 2003 and included in the Kyiv Resolution on Biodiversity; it is also highlighted by the EU as a key action to prevent the abandonment or intensification of these lands.

Agri-environment measures, an optional policy tool for farmers, provide compensation payments covering implementation costs and associated income losses to farmers who commit to preserving the environment and maintaining their farmlands through environmentally friendly practices for at least five years. Under the EU Common Agricultural Policy (CAP), Member States are obliged to co-fnance these measures: between 2007 and 2013, nearly 22 per cent of the expenditure on rural development, some US$27.3 billion (•20 billion), was devoted to them. Securing financial support and avoiding delay in payments is necessary to ensure farmer commitment.

In terms of biodiversity conservation, agri-environment measures are at their most successful over large areas, where they also contribute to the maintenance and enhancement of landscapes, protection of the historic environment and of natural resources, and the promotion of public access to the countryside. Their high costs, however, may limit their replicability in non-EU European and developing countries. Other limitations to their spread include potential loss of income for farmers and the difficulty predicting their effes on biodiversity.

Although forests currently cover 45 and 38 per cent of Europe's and the EU-27's territory respectively, only 26 and 4 per cent of these forests are considered to be undisturbed by humans. Most European forests are heavily exploited and the share of old-growth stands, crucial for forest species, is critically low. Nonetheless, Europe's total forest area is increasing thanks to national policy initiatives coordinated in the Forest Europe framework – a voluntary pan-European policy process for establishing sustainable management of the region's forests.

The Forest Europe process develops common strategies to meet challenges such as climate change and the protection of biodiversity and freshwater, both in Europe and globally. Since 1990, it has established a collaborative research network on forest ecosystems, a set of pan-European criteria and indicators for sustainable forest management, and a series of action programmes tackling cross-sectoral cooperation and national forest programmes. Sustainable forest management, as defned by the Ministerial Conference on the Protection of Forests in Europe, has been recognized as a commendable example of the ecosystem approach advocated by the Convention on Biological Diversity (CBD).

The benefits of Forest Europe include harmonization of forest policies in European countries that aim to achieve goals for protecting biodiversity, combating illegal logging and certifying carbon sequestration. Europe has gained 5.1 million hectares of forest since 2005, and between 2005 and 2010 about 870 million tonnes of CO_2 were removed annually from the atmosphere by photosynthesis and tree biomass growth, about half of it in the EU-27.

Eforts to enhance the sustainability of forests through management face a lack of national capacity and awareness, and intensifying competition in international forest product markets. There is therefore an urgent need for transnational coordination to address common and cross-border issues. National diferences also reffe the different roles of forests in various countries and the resulting political need to establish ofcial forest programmes.

The absence of a legally binding agreement on forests at a pan-European level cannot be considered a limitation to successful policy implementation, but at some point it could slow the process down, as common benchmarks and well-defned targets for evaluating effectiveness and efciency are lacking. In order to improve and accelerate the process, in June 2011 the Ministerial Conference on the Protection of Forests in Europe adopted the Oslo Ministerial Mandate for Negotiating a Legally Binding Agreement on Forests in Europe.

In European environmental governance, the integration of effective policies under multiple environmental themes and economic sectors is increasingly being taken into account. Even if such policies still have one environmental theme as an entry point, they increasingly cover wide ranges of related aspects. The recent 2009 EU Climate and Energy Package exemplifes such an integrated approach, including binding legislation to achieve three linked targets (the 20-20-20 targets).

It is through such integrated policies that multiple co-benefits can be obtained most cost effectively. Industrial CO_2 emission reduction through emissions trading, for example, will at the same time improve ambient air quality; and promoting renewable energy systems will not only reduce CO_2 emissions, but will also decentralize energy production, potentially improve energy security and provide employment opportunities and economic growth in small and medium-sized companies. Likewise, climate adaptation programmes will increase resilience to climate change effes such as fooding, drought, loss of biodiversity and increased vulnerability to disease, while at the same time improving ambient air quality and reducing greenhouse gas emissions, for example through adjusted agricultural practices, which will also contribute to more sustainable agriculture.

In summary, European examples of regional cooperation on the environment have served as a model for other countries and regions and can potentially serve in the future. Features include current formal institutional structures and the tradition of legislating to improve state and trends in various realms in an integrated way, though as contexts vary adjustments may need to be made in other parts of the world. Ongoing European attempts strive for continually improving environmental governance, underpinned by strong civil society participation and the recognized right of access to environmental information and justice in environmental matters as laid down in the Aarhus Convention, to date only applied in Europe. These efforts are essential for a proper and robust treatment of the shared environmental space and a healthy future for all.

References

COE (2000). *European Landscape Convention.* European Treaty Series No.176. Council of Europe,Strasbourg

EEA (2011a). *Greenhouse Gas Emissions in Europe: A Retrospective Trend Analysis for the Period 1990–2008.* EEA Report No 6/2011. European Environment Agency, Copenhagen

Ragwitz, M., Winkler J., Klessmann, C., Gephart, M. and Resch, G. (2012). *Recent Developments of Feed-in Systems in the EU – A Research Paper for the International Feed-In Cooperation.* A report commissioned by the Ministry for the Environment, Nature Conservation and Nuclear Safety (BMU), Bonn

UN (2000). *Millennium Development Goals.* United Nations http://www.un.org/millcnniumgoals/

Ziolkowska, J. (2009). Environmental benefit, side effects and objective-oriented financing of agri-environmental measures: case study of Poland. *International Journal of Economic Sciences and Applied Research* 2, 7188

11

Environmental Governance in Latin America and Carribean

The 33 countries of Latin America and the Caribbean vary significantly in size and economic development. The region includes both Brazil, the seventh largest economy in the world and small island developing states, with their open and fragile economies. Rich in natural resources, the region is home to approximately 23 per cent of the world's forests, 31 per cent of its freshwater resources and six of the world's 17 mega-diverse countries. Although these resources are not evenly distributed, the overall richness and economic importance of the region's ecosystems and its natural capital are undeniable.

Latin American and Caribbean countries face many challenges in managing their rich natural resources. Population growth, as well as unsustainable global and regional production and consumption patterns, drive the increasing demand for, and extraction of raw materials and other natural capital. This has led to the extensive conversion of natural environments to productive systems, with impacts on the region's biodiversity.

With 79 per cent of its population living in towns and cities, the region is one of the most urbanized in the world. It faces challenges in providing its burgeoning towns and cities with safe water and sanitation, and in addressing air pollution and the contamination of its freshwater, oceans and seas. The associated competition for scarce resources and the inequitable distribution of benefits have led to emerging socio-environmental conficts and risks to the traditional lifestyles and livelihoods of local and indigenous communities.

Global climate change exacerbates many of the region's existing problems. Extreme weather patterns and climatic events are increasing in frequency and intensity, and sea levels are rising. The impacts are already affeing the region's most vulnerable groups, including its small island developing states and many rural, indigenous and poor populations. Thus, it is ever more important to use water resources efciently and to conserve and sustain terrestrial, coastal and marine ecosystems. The challenge, however, is great, and the region is far from achieving some of the Millennium Development Goals (MDGs). Given the current situation, including poverty throughout Latin America and the Caribbean, there is an urgent need to implement more effective measures to halt and reverse the region's negative environmental trends.

The region has many laws relating to the environment but, at the same time, the lack of institutional management and capacity to implement and enforce them has constrained their effectiveness. In addition, policies are not keeping pace with production practices or adapting sufciently to global trends and integration. To address these challenges, governments need a stronger commitment to new policies and to making existing policies and mechanisms more effective.

Certain countries are progressing in incorporating new policy mechanisms, such as valuation of ecosystem services, payment for ecosystem services, climate-compatible development, innovative green fnancing mechanisms, and sound corporate practices, to name a few. Some progress is also being made in developing national environmental/sustainable development strategies that take both cross-sectoral and multi-stakeholder views into account. These positive lessons are a starting point for considering the options available to the region's policy makers.

It highlights policies considered to have the highest potential for increasing environmental sustainability and associated human well-being. A number of interrelated themes have been selected as priorities: environmental governance; water management; biodiversity; soil, land use, land degradation and desertification; and climate change. Sustainable management of oceans and seas is also important, especially to the region's small islands, so coastal and marine issues are also addressed.

Regional Context

Environmental governance has been identified as a priority theme for the region and is treated as cross-cutting with respect to the other selected environmental themes. This reffes the fact that sound environmental governance will ultimately reverse environmental degradation and help achieve the MDG targets and many multilateral environmental agreements.

Governance of the environment and natural resources in Latin America and the Caribbean is a complex mosaic. This stems from the wide diversity of governance systems with different degrees of institutional development and approaches to environmental issues, and different levels of governance mechanisms and performance. Regional and sub-regional mechanisms play an important role in environmental management, although in many cases the environment is not their main focus.

In recent decades, most Latin American and Caribbean countries have developed national environmental legal and institutional frameworks to formulate strategies and action plans for sustainable natural resource use and environmental protection. In addition, countries have begun to adopt a more cross-sectoral approach, with other agencies considering environmental issues in addition to those directly responsible for the environment. Despite these achievements, a limited capacity to implement and enforce existing legislation and poor institutional arrangements constrain effectiveness. The weak development of environmental policies in the face of economic, financial, commercial and technological globalization has aggravated the situation. Managing national environmental policies and balancing internal priorities among other sectoral needs, while engaging in multilateral efforts through multilateral environmental agreements, constitutes a major challenge for the region.

Another concern is policy and institutional continuity, which is especially important for environmental issues. The timescales over which policies, programmes and projects are realized do not always coincide with those of political terms of ofce. Options to strengthen the political authority of environmental agencies and maintain essential medium- to long-term efforts include longer terms of ofce and greater autonomy for technical environmental ofcers, and creative fnancing mechanisms to facilitate political independence.

Effective Environmental Governance

For effective and efcient functioning, a number of enabling conditions should support policy and institutional frameworks, including adequate financial resources, scientifc research and information, environmental education and a culture of environmental awareness. In addition, standard governance principles and values such as transparency, accountability, equity, sustainability and inclusive stakeholder participation should underpin any governance framework.

Policies for generating and disseminating information foster a better understanding of environmental conditions, problems and potential solutions and improve the science-policy interface. Reliable and timely information allows decision makers to respond appropriately and thus improves decision making. Where relevant, this information should also incorporate indigenous/ local knowledge. To influence policy and decision making effectively, environmental information should be transformed into easily understood, scientifically derived indicators to convey clear messages to policy makers and the public. Importantly, information should not be policy prescriptive but policy relevant, and should provide decision makers with alternatives and associated scenarios.

Relevant information and indicators also help in monitoring and evaluating the effectiveness of policies and determining if they have allowed management approaches to adapt to new conditions; these are important elements of good environmental governance. Good monitoring and evaluation programmes need to consider appropriate time frames and adequate baselines, and focus on results-based management using appropriate indicators. While well established in internationally sponsored projects, planned participatory monitoring and evaluation systems should also be used in government-run initiatives to quantify results and enable adaptive management. It is important that, in addition to quantitative scientifc information, monitoring and evaluation regimes include social, political and cultural qualitative data to assess results and develop methods to improve policy effectiveness. Indicators may be process-based to measure progress or outcome-based to measure effectiveness, and should include the evaluation criteria of coverage, effectiveness, sustainability and replication.

Environmental education gives people a greater sense of responsibility and increases awareness of the consequences of their actions. It promotes an environmentally conscious culture that helps to overcome a general lack of environmental awareness, one of the main causes of adverse change. Furthermore, an environmentally aware culture potentially improves public participation and increases public support for initiatives. For example, increased environmental awareness is credited with greater public support for developing payment for ecosystem services in Costa Rica.

Since the early 1990s, most countries have incorporated provisions for citizen participation in environmental legislation or in thematic or sectoral laws and have created a variety of citizen participation councils. Although national and local regulations have standards for public participation, including those for a variety of instruments such as public hearings and consultations, implementing them effectively continues to be a challenge.

Co-management is one of the most effective and efcient approaches to incorporating public interests in environmental decision making. The co-management of protected areas and watersheds by local communities, civil society organizations, indigenous peoples and even the private sector, has become a model of stakeholder participation. This approach has been widely adopted in such areas as biodiversity conservation and forest management. For example, public-private partnerships used in tandem with economic incentives to protect critical watersheds are evident in a number of countries.

In many cases, however, citizens are only consulted at the very end of the decision-making process. This has exacerbated confcts that integrated water resources management and multi-scale land-use planning are designed to prevent or resolve, including confcts over water resources and land tenure. It is increasingly clear that there is a need for mechanisms to ensure accountability and transparency to reduce the risk of corruption in decision-making processes and to increase financial fows to environmental programmes.

Negative externalities resulting from market forces are often considered a driving force of adverse environmental change. Thus, in developing future environmental policies in the region, it is of utmost importance to recognize the economic value of ecosystem services. An appreciation of the market value of ecosystem services, which reffes the link between the environment and human well-being in monetary terms, helps promote an environmental culture and improves the political viability of environmental protection. The use of economic incentives encourages citizens and businesses to make decisions based on the true long-term economic value of nature and the services it provides. Examples include Reducing Emissions from Deforestation and Forest Degradation and additionally for conserving and sustainably managing forests and enhancing forest carbon stocks (REDD+); payment for ecosystem services such as the Fund for the Protection of Water in Peru; and feed-in tarifs to support renewable energy. Valuing natural assets economically also allows decision makers in the public and private sectors to optimize their cost-benefit analyses and may be used to adjust national accounts and other economic indicators. Other tools, such as green funds and environmental taxes, can be used to raise funds for cash-strapped environmental agencies and causes. For example, the Trinidad and Tobago Green Fund couples both sets of tools to fund biodiversity preservation and ecosystem management.

Efective environmental governance, especially in the context of complex systems, requires that stakeholders collaborate and cooperate; it also requires coordination and harmonization of institutions, policies and other instruments. A number of platforms and mechanisms have been established to facilitate greater collaboration and coordination and improve coherence among governance systems, although these vary in nature, scale and level of success. One such mechanism is the Caribbean Sea Commission, which is one of several initiatives under way to strengthen the cohesiveness of the approximately 30 organizations involved from the sub-regional to the international level in coastal and marine management in the Caribbean Sea. A multi-scale governance framework is proposed for this large marine ecosystem by Fanning *et al.,* which accommodates the diversity of policy cycles at multiple levels and the links between them. Such a framework could be adapted for other ecosystems or environmental issues.

Environmental justice is "the fair treatment and meaningful involvement of all people regardless of race, colour, national origin, or income with respect to the development, implementation, and enforcement of environmental laws, regulations, and policies". In recent decades, several Latin American and Caribbean countries have made significant progress in environmental justice, particularly by enacting specialized procedures and mechanisms, as well as by enhancing the capacity of judiciaries; in some cases, this has included establishing specialized tribunals, for example the Tribunal Ambiental Administrativo in Costa Rica, and designating environmental prosecutors.

Although there are positive examples of judicial rulings in the region, there are still many challenges to improving environmental justice, including institutional and legislative weaknesses, low public participation and lack of awareness and information about people's environmental rights.

The emerging role of the judiciary is also important. In many countries, civil society organizations, prosecutors and individual citizens are using the judicial system to defend environmental rights. This occurs mostly through constitutional courts, but also in criminal and civil courts. In addition, the justice system has been proactive in resolving technically and legally complex disputes by overcoming procedural obstacles and adapting traditional legal institutions to the specifcs of environmental law. The judiciary still needs to develop a considerable amount of capacity in addressing environmental issues, however, particularly by training legal professionals, especially lawyers and prosecutors.

Environmental governance should be viewed as a cross-cutting theme across all the other priority issues identified in Latin America and the Caribbean. Despite its complex environmental governance mechanisms, the region has made significant progress in developing national environmental legal and institutional frameworks. Poor institutional arrangements and a limited capacity for implementation and enforcement, among other defciencies, however, have hampered their effectiveness. A number of enabling conditions need to accompany these frameworks including adequate financial resources, scientifc research and information dissemination, environmental education and an improved environmental culture. They also involve the standard governance principles and values of transparency, accountability, equity, sustainability and inclusive stakeholder participation. Such good governance can help reverse the current trend in environmental degradation and help to achieve the targets of the MDGs and many multilateral environmental agreements.

Integrated Water Resources Management

The availability of clean water in sufcient quantities and of sufcient quality was declared a human right by UN Decision 64/292 and is recognized in the constitutions of some Latin American and Caribbean countries. They were identified as potential options to address the Johannesburg Plan of Implementation Paragraph 26c, selected as the region's internationally agreed goal related to water.

Integrated water resources management has been widely acknowledged as a way of achieving long-term solutions to water problems because of its interdisciplinary approach. Its implementation in developing countries, however, has been rather slow. Integrated policies include those associated with:

- strengthened water governance;
- application of economic and financial instruments; and
- improvement of information on water quality and quantity.

Strengthened governance is both a cause and effect of an holistic view of water management because it implies a balance between public interest and the rights of the individual. Economic instruments and information are key tools in managing complex situations such as water scarcity, water-use conficts and pollution. Economic instruments include mechanisms to change the culture of water use, such as economic valuation and the polluter-pays principle. Information gathering, including outputs from indicators and monitoring processes, supports the management of supply and demand and also helps to sustain traditional knowledge about the links between water, people and the environment. Finally, in the context of climate change, water-related information systems to prevent disasters and manage risk are increasingly important for the region.

Integrated approaches to water management enable resources and capabilities to be used in an efcient, cost-effective and sustainable way, which is ever more important as demand for water increases with population growth, and as the impacts of climate change are felt. Other benefits include fewer water-related conficts, such as in managing transboundary basins and other competing uses; increased participation of stakeholders in decision making - including women, indigenous groups and other minorities - that can help reduce marginality and inequity and promotes transparency and accountability; increased water conservation and sustainable distribution; decision making and policy formulation based on evidence and traditional knowledge; and appropriate basin management that contributes to land-use planning policies, helps address issues of food security, ecosystem protection and waste management, and reduces transaction costs in water chains.

Integrated water resources management has only been implemented in Latin America and the Caribbean in a limited way due to fragmented and conficting institutional mandates, lack of skilled human resources, inadequate mechanisms for effective public participation, lack of sustainable fnancing and harmonization mechanisms, and a lack of structures and procedures to gather and present data.

Enabling conditions to promote integrated management include:

- water policy reform, including legislation and standards;
- water governance, including institutional frameworks to monitor and enforce legislation, development of institutional capacity to design and implement integrated management plans, projects and long-term programmes at different scales, and greater engagement and use of local knowledge through basin committees;
- land register development, stable governance arrangements, low transaction costs, credible enforcement arrangements and clearly defned rights and/or entitlements for land and water use;
- developing government capacity to collect tax revenues, so that funds can be efciently and equitably allocated to water programmes and projects; and
- education and information programmes.

Sustainable Water Provision and Consumption

The Latin America and Caribbean region has 31 per cent of the world's freshwater resources. However, given the region's rate of population growth, rapid urbanization and current patterns of water use, sustaining ecosystem services and an adequate water supply for future generations is an increasingly important issue. Investment in infrastructure is necessary, but it alone is not enough to solve the problem of water supply and demand. There is need for a change in policy making and management approaches, from those based exclusively on managing supply to the inclusion of both supply and demand management. Among users, there is need for a cultural change through education and economic incentives. Two main policy options may be considered:

- conservation and restoration of water-supplying ecosystems;
- promotion of water-use efciency in human consumption and production activities.

Ecosystems provide a wide range of services within a watershed. Thus, establishing and maintaining the minimum amount of water they require (environmental fows) is vital to ensure a balanced hydrological cycle and a constant water supply. In areas where resources are heavily exploited, improved water-use efciency is urgently needed through technological developments and by applying traditional and scientifc knowledge. This fosters measures to adapt to climate change as well as reducing costs for water users. Investments in water-use efciency include the control of unaccounted water at the grid level, installation of water-saving appliances, reuse and recycling systems, rainwater harvesting and water-saving irrigation systems, among others. Although initial investments are high, reduced water use translates into reduced costs in the long term.

Overall water policy needs reform to ensure that the policies proposed here do not remain isolated projects or campaigns, but have long-term effes. It is thus important to develop the political will to adopt legislation that will effect positive change through encouraging incentives and enforcing penalties. There must be economic incentives such as access to loans with low interest rates and equitable conditions, as well as water-efciency certification schemes. Management committees, civil society and multi-stakeholder participation are key to success. In summary, sustainable water supply and demand can be achieved when the economic, cultural and social value of water is acknowledged.

Expansion of Drinking Water and Sanitation Systems

To achieve MDG 7, 92.5 per cent of the population of Latin America and the Caribbean must have access to safe drinking water and 84.5 per cent to basic sanitation by the end of 2015. According to the most recent MDG report, the region has high rates of achievement for the first target and moderate ones for the second. This suggests that the sanitation target will not be met if prevailing trends persist. Furthermore, there are enormous diferences within segments of the population, between urban and rural areas, and between the three sub-regions.

The drinking water and sanitation policy cluster includes:

- freshwater augmentation;
- water quality improvement;
- wastewater treatment and reuse; and
- water conservation.

These policies are specifc to each sub-region on issues such as water use relative to water availability; existing water supply infrastructure including its condition and size, geographical watershed extent, number of people connected and number of people who receive measured water; user characteristics including socio-economic issues, consumption patterns, and essential and non-essential uses; and technical, financial and institutional resources. Examples of technological options to expand water availability are rainwater harvesting, water reuse, groundwater recharge and desalination.

These policies require a high level of commitment from governments as well as relatively high financial investments. In addition, maintenance costs, lack of technical competency - for example for desalination - and inefcient water-use habits could hinder the expansion of coverage. International cooperation is needed to fnance cases requiring special technical or social development skills that governments cannot aford.

Integrated Coastal Zone Management

Population density in the region's coastal zones is significantly greater than in inland areas. Coastal infrastructure, urbanization and tourism and land-based pollution are significant pressures on coastal and marine ecosystems. The rise in sea level due to climate change and the increasing frequency of El Niño/La Niña phenomena are also affeing coasts and changing coastline dynamics, ecosystem health, rainfall patterns and river fows, as well as damaging infrastructure.

Integrated coastal zone management is a multidisciplinary and intersectoral approach to land-use planning that promotes effective, meaningful and sustainable management of coastal resources. Similar to integrated water management, it assimilates the interests and needs of different stakeholders, maintaining ecosystems and their services in a cooperative and rational manner. In the Caribbean, for example, mechanisms have been implemented through the international project for Integrating Watershed and Coastal Area Management in the Small Island Developing States of the Caribbean (IWCAM) and the action plans of Barbados, Belize and Saint Lucia.

The coastal management policy cluster includes the establishment and execution of legislation, regulations, standards and procedures to prevent or minimize environmental degradation, and to protect and restore the quality and function of ecological systems within the coastal zone. It requires an appropriate legal framework, effective institutional structure, and information, data and knowledge for management. It also needs a clear and collectively recognized definition of the coastal zone's limits. The foundation for implementing this approach is a coastal zone management action plan, while strengthening monitoring and evaluation capabilities enables progress to be rigorously tracked.

Integrated coastal management promotes the preservation of ecologically sensitive areas such as mangroves, fosters the sustainability of important socio-economic activities such as fisheries and tourism, preserves natural ecosystem functions and services such as coral reefs, and improves the quality of the marine environment, for example by reducing contamination from vessels and in ports. Experience in Barbados, Colombia, Saint Lucia and the wider Caribbean demonstrates these benefits.

Biodiversity Management

Latin America and the Caribbean is home to approximately 70 per cent of the world's species and almost 20 per cent of its ecoregions. Its economy is highly dependent on this rich biodiversity, yet it is increasingly under threat from human activities. Although there are numerous biodiversity policies and measures in the region, collectively they do not effectively conserve its biological resources.

Addressing the driving forces that affect biodiversity requires equitable, evidence-based, participatory, cross-sectoral policies designed to protect and restore biological resources. In the context of the new Aichi Targets – 20 targets that form the framework for biodiversity conservation until 2020 under the Convention on Biological Diversity (CBD) – and given the region's biodiversity priorities, CBD Article 10 was selected as the internationally agreed biodiversity goal related to this priority issue. The following four policy options are considered able to help accelerate the region's progress towards meeting this goal.

Increasing and Expanding Protected Areas

Latin America and the Caribbean's protected areas, including marine, cover more than 500 million hectares in 4 400 different zones. They are considered to be one of the region's most important policy measures for conserving biological diversity. There is documented evidence that protected areas not only play a role in conserving species and habitats, but also deliver a range of ecosystem services and are considered important in climate change adaptation and mitigation. If properly managed, they can both contribute to national gross domestic product (GDP) and help to cover their own costs. Although not often realized, protected areas have the potential to provide a range of social benefits: improving equity and alleviating poverty as well as empowering women, communities and indigenous peoples – all of which are important considerations in the region.

Although protected areas have demonstrated both progress and success in biodiversity conservation in Latin America and the Caribbean, they face a number of challenges. An important one is that isolated areas often ofer insufcient protection, but creating biological corridors or improving landscape-scale connectivity can improve protected area performance. Greater connectivity can also improve species resilience to climate change and provide multiple benefits to humans.

Other ways to enhance protected area effectiveness in the region include:

- advancing conservation in marine and freshwater protected
- areas that are still largely under-represented;
- effectively integrating indigenous and local communities
- in protected area management, including, where relevant,
- by promoting indigenous- and community-conserved areas;
- promoting the links between conservation and development goals, using land-use planning as a fundamental tool; improving research capability and strengthening links between research and decision-making frameworks; and
- strengthening the capacity for managing protected areas

In addition, key instruments for protected area management include ecotourism and sustainable tourism programmes; balancing the relationship between conservation and development through mechanisms such as payment for ecosystem services, including for carbon dioxide (CO_2) capture and sequestration services and environmental stewardship and usage fees; and the selective extraction of resources. Measures such as tax incentives, preservation easements, education, decentralized administration, partnerships with international organizations and outright land purchases may also encourage and promote protected areas and associated corridors and linked landscapes.

Applying the Ecosystem Approach to Biodiversity Management

The ecosystem, or ecosystem-based, approach is increasingly recognized as an important strategy in biodiversity management, especially in the context of climate change. According to the CBD, it is "a strategy for the integrated management of land, water and living resources that promotes conservation and sustainable use in an equitable way".

The ecosystem approach is not designed to replace other management and conservation approaches, but rather to complement and support them, for example sustainable forest management, integrated river basin management, integrated marine and coastal area management and sustainable fisheries. In addition, approaches such as creating protected areas, corridors or biosphere reserves and species conservation programmes, as well as action under existing national policy and legislative frameworks, can be integrated to deal with complex ecological situations.

The ecosystem approach has been identified as a key policy in Latin America and the Caribbean for two main reasons: it is useful for managing water resources, wetlands and land, and in developing payment for ecosystem services; and many pristine ecosystems still exist with high conservation value. Because of their size, the Caribbean's small island states also present excellent opportunities for implementing the ecosystem approach, and could serve as case studies for ascertaining its strengths and weaknesses.

Although there are several on-the-ground initiatives applying an ecosystem approach in the region, this has often been done on an *ad hoc,* single-project basis, which remains a challenge. Such initiatives need to be better integrated into institutions, including those concerned with sectors outside biodiversity conservation, such as agriculture, fisheries, forestry and health. More research is also needed to support the development of a monitoring and evaluation framework for each of the principles of the ecosystem approach. In addition, issues such as illiteracy, land boundaries and the cost of participatory processes all need to be considered in integrating and assessing the impact of the approach in Latin America and the Caribbean.

Enhancing Biodiversity Conservation

A number of options grounded in economic theory present promising opportunities for both mainstreaming biodiversity issues and reducing driving forces, while simultaneously supporting development processes and promoting human well-being. Among these is the payment for ecosystem services – or PES – mechanism, which was largely pioneered in Latin America and the Caribbean, and which is gaining popularity worldwide as an effective approach to dealing with biodiversity loss.

In general terms, PES schemes or systems ofer incentives, usually monetary ones, to individuals to protect and ensure the delivery of key ecosystem services at local, national and regional levels. The mechanism can address many of the driving forces of biodiversity loss in the region, especially habitat loss and unsustainable land management, as it usually aims to protect and/or rehabilitate natural vegetation. In addition it can support many existing policies.

Monetary compensation provides a tangible incentive to protect habitats and their biodiversity by providing sustainable livelihoods; it also mitigates the initial needs that drive unsustainable biodiversity resource use. As such, PES has the potential to increase employment and equity. This reduces poverty, since low-income groups and ecologically sensitive land in the region's developing countries often co-exist. Given that there is a strong link between habitat protection, rehabilitation and a number of ecosystem services – such as water provision and purification, coastal protection, mitigation of greenhouse gas emissions and protection against soil erosion – PES schemes bring multiple co-benefits to a range of sectors.

Payment for ecosystem services is not without challenges. Its limited application and a lack of information on economic valuation highlight the need to invest more in research and furthering the scientifc understanding of local environmental conditions. Certain services cannot be measured, however, and determining the seller of these services is also difficult. Moreover, finding buyers and mobilizing funding is the greatest challenge to implementing PES.

Coupling PES with innovative fnancing mechanisms, however, could address this. Examples include ring-fencing budget allocations for environmental protection, as in the Programme for Forest incentives in Guatemala; earmarking government taxes for environmental protection, such as Brazil's ecological value added tax; providing environmental funds like Trinidad and Tobago's Green Fund; and setting up public-private partnerships.

Access and Benefit Sharing

Latin America and the Caribbean's rich genetic resources are important to local communities in sustaining their livelihoods, and especially in providing food security. However, many genetic resources are also the basis of commercial use and production. To promote equity and safeguard the genetic diversity and associated traditional/local knowledge within the region's countries, there has been a growing interest in access and benefit sharing.

Argentina, Brazil, Costa Rica, Mexico and Peru have developed access and benefit-sharing legislation at the national level and the Andean Community states and Central American Commission on Environment and Development have done so at the sub-regional level. At the national level, there are two main groups of associated legislation: framework laws on sustainable development, nature conservation and biodiversity; and dedicated or stand-alone national laws or decrees (Brazil). Access and benefit-sharing considerations can also be incorporated into general environmental framework laws, or existing laws and regulations can be modified to address them, although this has not yet occurred in Latin America and the Caribbean.

The CBD's Nagoya Protocol on Access to Genetic Resources and the Fair and Equitable Sharing of Benefits Arising from their Utilization, adopted in October 2010, now provides a global framework for improving legal certainty and transparency related to access and benefit sharing, and could help Latin American and Caribbean countries overcome various problems in implementing relevant policy. As of April 2012, 14 countries in the region are signatories to the protocol. To maximize the advantages of access and benefit-sharing policies, attention needs to be paid to several factors, including:

- undertaking research to better understand and apply access and benefit-sharing principles and the Nagoya Protocol in the regional context;
- enhancing human, technical and financial capacities;
- clarifying legal defnitions and interpretations;
- understanding and dealing with the transboundary nature of genetic resources;
- protecting traditional knowledge; and
- negotiating tangible benefits instead of focusing only on access procedures.

Land Use, Land Degradation and Desertification

Pressure on land resources has increased in recent years despite international goals to improve land management. To halt and reverse land degradation and ensure renewable resources are used sustainably, policies allowing productive activities with minimal impact on natural ecosystems and their environmental services are required. These include land-use policies that prevent inefcient or inappropriate transformation for agriculture, livestock or illegal crops.

Examples include sustainable forest management, increasing efciency and intensifying productivity to reduce environmental impacts, improving waste management, decreasing the amount of new land being cultivated, and helping prevent confcts due to land, water and other resource shortages. At the same time, environmentally friendly productive activities with tangible economic benefits for land owners and environmental services for society need promotion. Finally, degraded ecosystems need to be rehabilitated and their sustainability ensured to restore the productivity chain that supports ecological balance and social and economic well-being.

Multi-scale Land-use Planning

Land-use plans take account of all the resources and dimensions involved in the development process and help implement integrated land-use management, water resource planning and conservation priorities, while also encouraging inclusive, multi-stakeholder participation. Land-use planning considers a number of inseparable elements: land, renewable and non-renewable resources, and a coherent view of current and historic lands and their uses, existing services, accessibility and cultural influences. Land uses such as agriculture, agroforestry, livestock production, industrial development and mining, among others, must also be considered.

Land-use policies involving stakeholder participation, regulations and financial instruments are necessary to prevent a number of land-related confcts, including transboundary controversies over scarce resources between and within sectors and countries; land tenure and titling issues for rural families – where creating cadastral and registration agencies helps create stability; and the rights of minority groups such as indigenous communities and women to land ownership. Furthermore, land-use planning can be an effective mechanism to prevent resource depletion and environmental degradation.

In a wider sense, land-use planning should also include marine and coastal zones because of the interaction between land and aquatic environments. According to UNEP-CEP, land-based activities are the greatest threat to Caribbean coastal and marine habitats. Likewise, the integrity of marine and coastal ecosystems, which is linked to terrestrial well-being, also affes social resilience, especially in terms of public health and livelihoods.

Several countries have either implemented or are preparing land administration plans, including national coastal management acts. Land administration projects in Latin America are focused primarily on facilitating a land market. Although projects have social equity and environmental sustainability goals, these are largely of secondary importance. In some countries, for example Bolivia, Ecuador and Peru, advances in land administration require improvement in the property market infrastructure. Land tenure also needs to be stabilized, especially in post-confict situations such as those of Colombia, El Salvador, Guatemala and Nicaragua.

There are several challenges associated with land-use planning, including the lengthy process of collecting land-use and cadastral data, which requires information about legal titling that is hampered by legal barriers. The often illegal nature of historic land tenure processes, including forced displacement of peasants resulting from civil confict or corrupt but sophisticated schemes, is another constraint. In addition, transaction costs may be a major obstacle to registration, particularly for the poor. Finally, land-use planning policies may discriminate against minority groups such as indigenous and peasant communities, since many land administration projects are based on a simple territorial demarcation and a title issued in the name of the group.

The ceding of subsoil rights in indigenous territories to such outside economic interests as oil and mining companies can result in major physical intrusions and habitat damage, generated, for example, by the construction of infrastructure and roads. The Xingú River basin in Brazil is an example of the successful protection of indigenous territory from deforestation through land policies that include community participation.

Regional experiences show that it is much more important to obtain general agreements on land policy direction than to require, *a priori*, a technically perfect legal framework. New legal frameworks have proven ineffective, as insufcient attention has been paid to stakeholder discussions or the dissemination of their rights.

Sustainable Agriculture and Livestock Production

In land-use policy making, it is necessary to distinguish between small-scale and large-scale commercial agriculture. Out migration and land sparing, a system under which some land is farmed intensively to maximize yields while other land is protected as a nature reserve, allow more land to be devoted to preserving biodiversity and providing ecosystem services, but small-scale agro-ecological systems appear to be a good option for combining hunger alleviation and biodiversity preservation. Perfecto and Vandermeer suggest using a policy-making matrix that integrates agricultural and conservation elements to boost small-scale agro-ecological options. Policy-making matrices that use a framework built around payment for ecosystem services can significantly strengthen this approach.

Policies that promote organic agriculture, silvo-pastoral practices, ecotourism and sustainable rural tourism, fall within this category. Silvo-pastoral strategies such as planting trees and shrubs in pastures, fodder banks or trees and shrubs as hedges, induce farmers to increase practices that provide ecosystem services – improving biodiversity, sequestering carbon and conserving water resources. Policies fostering ecotourism including sustainable rural tourism promote the optimal use of natural resources and respect for socio-cultural diversity, which improves economic viability and distributes benefits more equitably. Well planned rural tourism can promote social development and equity, providing more opportunities for vulnerable groups such as youth, women and indigenous communities.

Examples of successful land-use planning are agritourism initiatives in the Caribbean; Cuba's transition to organic agriculture; silvo-pastoral practices and PES in Colombia, Costa Rica and Nicaragua; and rural-based community tourism in Guatemala and Nicaragua.

Land-use policies for large-scale commercial agriculture, which occurs in Argentina and Brazil, should promote sustainability through integrating existing knowledge with input-based farming technology. Policy options include the adoption of agronomic practices such as zero-tillage, minimum tillage, crop diversification, crop rotation and integrated pest management, combined with strategic applications of fertilizers and irrigation water, using low-impact pesticides and the expansion of precision farming procedures.

These practices have had positive impacts in Argentina, where public-private partnerships have been successful; in Paraguay's poultry industry where initiatives for cleaner production have been effective; and in Uruguay's environmentally friendly rice cultivation.

Empirical evidence in Latin America and the Caribbean suggests two ways of developing environmentally friendly livestock production systems, regardless of the farming scale: first, by increasing beef productivity through the dilution of maintenance costs; and second, by integrating crops, pastures, fodder and livestock production. The first case results in a significant reduction in land, water, fossil fuels, feed consumption and outputs of manure and greenhouse gases. In the second case, experience with integrated crop rotation, livestock production and zero-tillage operations in the Brazilian Cerrado allowed grain and meat to be produced sustainably on the same lands, thus eliminating the need to deforest more land.

There are examples of successful organic agriculture in most Latin American and Caribbean countries, although there is a need to harmonize policies, particularly those related to market access and distribution. Many countries are establishing regulations and standards for organic production while a few are providing limited financial support to pay certification costs during the conversion period. The current global market for organic production has encouraged the development of standards, certification processes and public-private partnerships to facilitate market access to organic produce.

Access to micro and small-scale credit in poor rural communities is necessary to ensure that land use is managed sustainably. The enabling conditions that facilitate the expansion of sustainable models of large-scale commercial farming generally rely on access to modern technology, for example precision and low-impact farming and information and communication technologies; updated agronomic knowledge; the professional capacity of farmers; good international prices; the financial capacity of individual farmers and investment funds; and credits to farmers' cooperatives.

RESTORATION OF DEGRADED LANDS

In addition to impacts on biodiversity and the economy, land degradation has social consequences. These include increased vulnerability to foods and dust storms; health risks such as from vector-borne diseases associated with deforestation, and illnesses from contaminated sites; loss of environmental services including water source recharge; and decreased carbon sequestration and evapotranspiration. Thus, the region should prioritize the restoration of degraded lands, which complements conservation and ecosystem management policies oriented to climate change mitigation and adaptation, reduces disaster risk, and helps maintain the hydrologic cycle and water sources.

All available land, particularly degraded or marginal areas, needs to be used efciently to satisfy the socio-economic and environmental needs of current and future populations as well as to conserve natural ecosystems. Given the environmental, social and economic benefits of land, it is important to institute restoration or rehabilitation policies. Land can simultaneously generate profits through agriculture, livestock husbandry or forestry, and maintain and purify water sources, reduce the risk of foods and mudslides and improve people's living conditions. Given the high costs associated with restoration projects, better economic instruments are needed, including government commitments that promote and fnance these projects.

Restoring land and environmental services provides new options for productive activities, reduces the vulnerability of populations and decreases the conversion of natural ecosystems to agriculture or pasture. Other commercial activities, such as ecotourism, can also be encouraged. In addition to reclaiming soil and accelerating forest regeneration, for example, the biological corridor project of Nogal-La Selva in Costa Rica represented an economic incentive for local farmers, while the reforestation project in the Panama Canal watershed reduced the costs of maintaining the canal's infrastructure. Restoring degraded lands benefits both market-based and non-market-based ecosystem services at multiple spatial scales.

Land restoration policies and action account for the specifc conditions of the site and the benefits expected. Efective restoration requires setting specifc and clear goals as part of the planning process and ensuring that the parties compromised by the recovery of degraded lands accept them. Thus, implementing the policies requires effective participatory mechanisms that include indigenous and other disadvantaged groups. For this reason, certainty and legitimacy of land tenure are also required.

Climate Change

Climate change is exacerbating many of Latin America and the Caribbean's environmental challenges; it also threatens development gains, poverty reduction and economic growth. Although the region accounts for a relatively modest 12 per cent of the world's greenhouse gas emissions, it is already experiencing the adverse consequences of climate change and variability. As vulnerability to climate impacts increases, addressing the underlying drivers of risk becomes a top priority. Poverty, marginalization, exclusion from decision-making processes, lack of opportunities, limited access to credit, inadequate education, poor basic infrastructure, inequity, insecure land tenure, and other factors external and internal to the region, continue to exacerbate its vulnerability.

To address climate change, the region needs to commit to the sustained implementation of international and regional agendas, such as the United Nations Framework Convention on Climate Change and its Kyoto Protocol, and the Hyogo Framework for Action. It should also commit to the sustainable environmental management of forests and key ecosystems; energy efciency and the development of new, renewable energy sources; ecoagriculture; and the transformation of transport systems, implemented in a socially and environmentally responsible way by respecting people's and communities' rights, and supported by international financial and economic mechanisms.

With the world's highest percentage of urban dwellers, the Latin America and Caribbean region faces many climate change challenges in its large and growing cities, many of which are located in higher-risk areas on low-lying coastal plains. To build resilience among the segments of the population most in need, municipal policies should be city-specifc and work in tandem with national and international efforts for mitigation and adaptation.

Although the region's cities have taken many initiatives on policies and activities both to mitigate and adapt to climate change, these have focused mostly on the former. It has been difficult to promote adaptation at the local level without the necessary support from higher levels of government and the international community. This has left a gap in the support and funding of locally determined, locally driven adaptation efforts that serve and work with those most at risk. The best opportunities to adapt to climate change are linked with action to address the underlying causes of vulnerability and respond to more than one problem at a time.

Reducing the Vulnerability of Populations

Implementing adaptation measures that consider economic, social-ecological and political criteria is an immense challenge. Fostering research programmes on the impacts of climate change, deforestation and land-use change on the natural environment and the social fabric are priorities, as are strengthening evidence-based policy making and the appropriate institutional infrastructure.

Policies for adapting to climate change are critical for strengthening natural capital management. This is especially the case for managing changing water fows and improving ecosystem resilience, strengthening direct protection against climate-related threats in cases for which collective action is needed, and strengthening technology transfer and knowledge fows.

The following presents a more detailed analysis of the many issues related to adaptation policy development in Latin America and the Caribbean, structured as four policy groupings.

Strengthening ecosystem management for improving resilience: some countries have made significant efforts to provide a more solid methodological and analytical evidence base for understanding the relationship between ecosystem health, resilience and vulnerability. They have also developed economic cost-benefit analyses of ecosystem policy options and their potential in reducing the vulnerability of societies. Innovative policies and financial mechanisms for delivery are also required, as are sustainably resourced and multi-stakeholder capacity building and active participation of local stakeholders in implementing the process. Land-use planning and protected areas are local mechanisms for managing ecosystem services that include the concept of risk reduction.

Towards resilient infrastructure: in light of the risks posed by extreme weather events, reducing the vulnerability of infrastructure systems should be a central objective of climate change adaptation policy. The region has a wide range of potential policy instruments addressing these concerns, the most cost effective and efcient of which rely on enforcing sustainable building standards and relocating vulnerable populations. Large-scale projects to build or replace infrastructure in the coming years present a tremendous opportunity to ensure that physical infrastructure and land-use systems are resilient in a changing climate. The Barbados boardwalk is an example.There is also significant opportunity to improve the cost effectiveness and sustainability of climate-resilient infrastructure investments by more systematically considering ecosystem-based approaches as a component of comprehensive infrastructure adaptation strategies. Another strategy is to integrate disaster risk reduction concepts and methodologies in public investments, as the governments of both Peru and Costa Rica have done.

Strengthening weather monitoring and forecasting tools: early warning systems, one of the main branches of disaster risk reduction, include the monitoring and forecasting of impending events. A number of key intergovernmental organizations work to further early warning policy at the sub-regional level under the Hyogo Framework for Action and through the Regional Platform for Disaster Risk Reduction. Among them are the following: the Centre for Natural Disaster Prevention and Coordination (CEPREDENAC) in Central America; the Andean Committee for Disaster Awareness and Prevention (CAPRADE); and the Caribbean Disaster Emergency Management Agency (CDEMA); as well as humanitarian networks such as the recent REDHU (Humanitarian Assistance of MERCOSUR). Cuba, Mexico, Central America and the Caribbean small islands have implemented weather monitoring and forecasting tools that protect populations from injury and disease.Although the region's early warning systems lessen the loss of life, decrease injuries and mitigate against property damage, the World Meteorological Organization stresses the need to re-evaluate national and local emergency preparedness and response plans, which should be based on hazard and vulnerability mapping. It also stresses that countries should strengthen their monitoring and forecasting infrastructure and the skills of technical agencies while improving access to data and technology; strengthen dissemination channels that link national early warning systems to communities focusing on cultural and community needs; and address sustainability issues on the basis of available resources.

Adaptation policies for social resilience: decreasing vulnerability while increasing resilience is central to development, environmental sustainability, climate change adaptation and disaster risk reduction. Policy efforts can be integrated around this central challenge. Climate change adaptation policies based on social inclusion embrace challenges and opportunities associated with addressing the needs of all segments of the region's population. They are particularly sensitive to the most vulnerable, such as the rural and urban poor and indigenous peoples with traditional lifestyles.

The region's rural households depend heavily on agriculture. Thus, adaptation strategies for coping with the impacts of climate change on agricultural productivity and food security among poor rural households need to include access to such key elements as land, labour, fertilizers, irrigation, infrastructure and financial services as well as technological alternatives. Examples of good policy instruments are the region's agroforestry systems and the Rainforest Alliance's Climate Module, which fosters the adoption of good agricultural practices to reduce greenhouse gas emissions and to enhance the capacity of farms to adapt to climate change in Central America.

Households, communities and the larger society are increasingly adopting approaches that protect them against the negative impacts of climate change. These include good public policies, such as providing public health services, education, social protection schemes, and supporting active and efcient civil society organizations or government agencies, a solid and well maintained infrastructure, good governance and healthy public fnances.

There are examples of ecologically oriented social policy in Brazil, including Bolsa Verde, which provides funds to the very poor who work towards environmental conservation, and the State of Amazonas Bolsa Floresta programme. Other examples of policies that increase social resilience come from Bolivia, Colombia, Nicaragua and Peru.

Sustainable Forest Management

Maintaining existing forests can be one of the most efcient and cost-effective options for mitigating CO_2 emissions, as can be seen in Brazil, Central America and Mexico. Protecting and restoring native forests – vitally important in sustaining the livelihoods and cultural heritage of many Latin American and Caribbean peoples – is promoted through such sustainable forest management strategies and results-based payment schemes as REDD+ or the recently created Amazon Fund in Brazil. Such strategies should focus on carefully integrating and providing benefits to rural and indigenous communities, as there are strong potential synergies with efforts to protect and rehabilitate forest resources. Policy action can rely on a variety of instruments, including payment for ecosystem services, public-and private-sector engagement or command and control, as such approaches can make REDD+ more effective in curbing greenhouse gas emissions. Peru's Forest Conservation Programme, Bolsa Verde of Brazil, as well as environmental services certificates, payment for ecosystem services and forest credits in

Costa Rica (FONAFIFO programme) are examples of the region's forest conservation policy instruments.

Successful efforts at knowledge generation, long-term policies on forest management, and native forest protection and restoration schemes typically strive to provide better information on the value of forest functions and products; strengthen multi-stakeholder involvement; create stronger links between legal, social, environmental, economic and technological tools; and critically evaluate the effectiveness of their objectives by continuously monitoring greenhouse gas emissions reduction and local sustainable development.

Encouraging Diversification of The Energy Matrix

International prices play a decisive role in defning Latin America and the Caribbean's policies related to fossil fuels. Renewable energy sources have been developed to address growing energy needs, with hydroelectric projects the preferred energy investment.

Renewable energy sources are a positive alternative to fossil fuels; nevertheless, renewable energy projects can affect the environment and the livelihoods of local communities and, as a result, need to be planned carefully. Given the region's diverse potential for renewable energy - biomass, solar, wind, wave and geothermal - the policy cluster proposes the introduction of renewable sources to the energy matrix.

The benefits of renewable energy sources include:

- the decentralization of investment towards less developed regions, which helps create jobs (for qualified personnel), capacity building and technology transfer;
- a cost-effective alternative to expensive grid extensions; and
- great potential to lower emission costs effectively, thereby reducing energy dependence and positively affeing trade balances.

The suggested policies combine long-term feed-in tarifs with subsidies and tax incentives as ways to provide investment and financial support for the supply chain of electricity generated from renewables, including the transformation of raw materials, manufacturing, and installation of components and systems.

Depending on the policy and regulatory mechanism, renewable energy may increase consumers' energy costs in the short term. However, income-specifc tarifs fnanced by reallocating counterproductive subsidies for non-renewable sources can often help to balance this distortion. If, however, oil prices fall, the opportunity cost may decrease to levels that might not cover costs. This can be addressed by introducing quota-based incentive programmes and long-term contracts with stable prices. Adopting policies that subsidize the use of renewable sources in terms of installed capacity (kilowatts), or pay per kilowatt hour generated and sold, can help to improve renewable energy. Likewise, mechanisms such as green certificates, research and development subsides, internalization of external costs and environmental taxes can foster an increase in the share of renewable energy sources in the energy matrix.

Policies involving smart grids and decentralized power generation have the potential to promote greater generation, transport and distribution efciency and to simultaneously scale up renewables, specifically solar and biomass. Complementarily, transboundary cooperation and integration in the energy sector have been shown to increase electricity supply, widen coverage and enhance system functionality across the region. The Renewable Energy Observatory of Latin America, the Energy and Climate Change Partnership for the Americas and the Mesoamerican Electric Interconnection are examples of policy strategies related to regional cooperation in the energy sector. The Tender System for Alternative Renewable Energies in Brazil and the Energy Strategy Guidelines from Uruguay encourage diversification of the energy matrix.

Enhancing Efficiency and Low-carbon Mobility

These policy options aim to reduce energy demand in the residential sector and transport systems while providing more effective and expanded energy distribution to the population. Financial instruments, such as cap-and-trade systems and carbon taxes, funds for research and development and compliance instruments might be adopted as part of the same strategy. A proposed reorganization of public transport systems would improve the efciency of fossil-fuel use and road space, and change the paradigm from individual and private to public and inclusive.

Related policy strategies foster the use of minimum standards of energy efciency for electrical appliances (lighting, cooling and heating) and individual vehicles (fuel efciency standards and the promotion of hybrid cars); and the adoption of energy efciency stamp programmes and specifc nationally appropriate mitigation action. To this end, it is essential to combine public fnancing instruments, market initiatives and specifc policies for research and development, as well as technology transfer to enhance the international transfer of resources associated with new technologies.

The main benefits of these strategies will be realized over the long term. Some studies reveal that energy efciency policies have usually reduced implementation costs. Furthermore, these policies can help reduce the negative impacts on human health by improving air quality; decrease external energy dependency; increase the reliability of power supply; control demand growth with the potential to reduce energy consumption by 20–25 per cent; increase productivity and employment; increase the efciency and competitiveness of domestic energy-intensive industry; and diminish congestion in cities.

In conjunction with its residential energy efciency policies, Latin America and the Caribbean has demonstrated the potential to further expand the green design and building market, especially for social housing. One exemplary effort is the government of Mexico's This is Your House initiative and the National Housing Commission's associated Green Mortgage Programme (Comisión Nacional de Vivienda or CONAVI). The government of Brazil is developing initiatives in this vein through the Ministry of Cities' planning instruments within the framework of its Multiannual Plan for Decent Living.

Policy Options and Environmental Priorities

To address the complexity of environmental issues, environmental policy making is evolving to transcend traditional, compartmentalized approaches by becoming more integrated and cross-sectoral in nature. It has presented clusters of policies that are believed to have strong co-benefits. An assessment of the policies in the clusters deemed them to be the most co-dependent and mutually supportive policies required to achieve the internationally agreed goals chosen for each thematic issue.

Moreover, in a number of instances the policies or policy clusters associated with certain themes were underscored as benefiting, or being strongly linked with, other policies and environmental themes. Apart from being advantageous to the environment, these policies have positive socio-economic and political impacts. In addition to biodiversity management, policies promoting payment for ecosystem services, for example, are used across a number of issues and in almost all sectors, including land, water and climate change. Policies that focus on integrated management of water resources or the ecosystem approach can also benefit other sectors such as agriculture, fisheries, forestry and land. Many of the climate change policies will ultimately co-benefit the management of land, water and biodiversity resources. Policy makers may find that understanding the links and co-benefits is useful in determining how to maximize the efciency of existing policies or measures, and in prioritizing the development and implementation of new ones.

Latin America and the Caribbean's ecosystems and associated natural capital are important to both the region's countries and to the entire planet. However, persistent negative environmental and related socio-economic trends are a clear indication that the measures so far established and implemented to protect them - at national, sub-national or supra-national levels - are insufcient to address either the rate or scale of conversion and consumption prevalent within the region. As a result, Latin American and Caribbean countries continue to face such issues as poverty, inequity and social confict related to environmental quality.

The consideration has been given to policies, approaches and instruments that have demonstrated the potential to improve sustainability in the region, especially for the issues deemed of highest regional priority.

The most salient point arising from the assessment of policy options is that strong environmental governance is a cornerstone for ensuring the success of policies geared at improving sustainability. Without strong governance frameworks to support environmental decision making, efforts to ensure greater environmental sustainability are unlikely to be effective. The following factors were identified as fundamental for strengthening governance frameworks:

- adequate financial resources;
- access to scientifc research and information;
- environmental education and the development of an environmental culture;
- the standard governance principles and values of transparency, accountability, equity, sustainability and inclusive participation of all stakeholders; and
- continuity in political systems.

The current limited impact of policies addressing environmental trends also highlights the need to emphasize the root causes that drive change throughout the region. Too often, policies tend to focus on the direct pressures affeing ecosystems and their services, because these are the best understood or they are easiest to deal with. There is, thus, a need to invest more in understanding these drivers and the ways they work together. Greater integration of environmental considerations into broader development processes is also needed.

The thematic issues covered have highlighted the interconnectedness of, and links between and among environmental issues. Most of the policy clusters are likely to benefit multiple sectors, once properly implemented. Thus, a close examination of cross-sectoral benefits is an important strategy for policy-makers to apply when considering the priorities and trade-offs associated with implementing a policy or a cluster of policies.

It suggests that existing policies, mechanisms and institutional frameworks at the sub-national, national and regional levels in Latin America and the Caribbean ofer a good starting point for strengthening environmental management. In many instances, it is not necessary to reinvent policies and their implementation, or to continue adding to the already saturated landscape of policies. Rather, what is required is a closer examination of existing policies and institutions to see how better to enable and strengthen them to serve more effectively. This approach could help to circumvent the long, sometimes onerous, processes required to build policies and/or new institutions from the ground up, and could accelerate the rate at which countries can work towards meeting internationally agreed goals.

Finally, cooperation is an important element in improving sustainability in the region. Cooperation between and among its countries would facilitate the sharing of information, expertise and technology transfer – the lack of which may currently limit countries in moving to more sustainable paths of development. It could also help to improve the management of ecosystems and species, which commonly cross national boundaries. Cooperation at a global level is also important to ensure that the region's natural capital is maintained and shared in a sustainable and equitable manner.

References

Andrade Perez, A. (ed.) (2008). *Applying the Ecosystem Approach in Latin America.* (translator Medina, M.E.). IUCN, Gland

Bennett, A.F. (2003). *Linkages in the Landscape: The Role of Corridors and Connectivity in Wildlife Conservation.* Second edition. IUCN, Gland, Switzerland and Cambridge

CBD (2004). *The Ecosystem Approach (CBD Guidelines).* Convention on Biological Diversity, Montreal

Ferraro, P. (2001). Global habitat protection: limitations of development interventions and a role for conservation performance payments. *Conservation Biology* 15, 9901000

Forero, E.G. (2008). The EA and water management: a Latin American perspective. In *Applying the Ecosystem Approach in Latin America* (ed. Andrade Perez, A.) (translator Medina, M.E.). IUCN, Gland

Maretti, C.C. (2003). *Protected Areas and Indigenous and Local Communities in Brazil.* WCPA Ecosystems, Protected Areas and People (EPP) project. IUCN, Gland

May, P. and Millikan, B. (2010). *The Context of REDD+ in Brazil: Drivers, Agents and Institutions*. Center for International Forestry Research (CIFOR), Bogor

12

Policy Approaches to Environmental Governance in North America

Environmental governance in North America is best characterized as multi-faceted, partly reffeing the nature of the federal political systems, ideological flux, evolving socio-economic constraints, and the dynamics of environmental issues as well as the knowledge associated with them. Federal governments are no longer the primary leaders in setting the policy agenda or devising innovative policy instruments, yet they remain essential to the ultimate success of those policies, help ensure harmonization across jurisdictions and prevent the development of environmental inequities. In addition, there is a strong tendency to favour market-based instruments because of early successes, and to overlook traditional regulatory instruments. Finally, relative federal disengagement has opened the door to policy initiatives and innovations at the sub-national levels of states and provinces or municipalities, as well as to regional transborder cooperation. The latter is extensive and continues to expand, and its dynamics are further supported by the Commission for Environmental Cooperation, which oversees the environmental accord of the North American Free Trade Agreement (NAFTA).

North America has used a variety of policy approaches to environmental governance, beginning with regulatory policies, then gradually developing market mechanisms, complemented by measures designed to improve accountability and transparency. The region was a pioneer in cross-border governance, which dates back at least to the 1909 Boundary Waters Treaty, and in developing international environmental law and national parks, including cross-border parks. In the last 20 years, this governance has deepened cross-border ties through the creation of the Conference of New England Governors/Eastern Canadian Premiers on climate change and the Commission for Environmental Cooperation, and by reinforcing cooperation between provinces and states in managing the Great Lakes and St Lawrence River, as well as on a variety of other issues, notably protection of waterbirds and sea mammals. The Georgia Basin/Puget Sound International Airshed Strategy in British Columbia and Washington State, for example, is currently the most active bilateral arrangement regarding air quality. For their part, the proposed creation of watershed boards across the entire Canadian/US border would represent a major leap in the International Joint Commission's regulatory potential. Canada and the United States have also established several jointly protected areas that further harmonize policies.

North America has pioneered the use of many market instruments, now being used with increasing frequency, and there is evidence that some have succeeded in changing behaviour. Command-and-control mechanisms, however, still form the backbone of environmental policy. Because of recent improvements in measures designed to foster accountability and transparency, these increasingly used instruments strengthen the effectiveness of both market instruments and command-and-control mechanisms. Rarely are any of these used exclusively to address a particular environmental issue; it is more common to see a variety of instruments applied. For example, to address littering, many North American municipalities and states or provinces have laws that require a deposit on bottles and cans. This deposit provides a financial incentive – a market instrument – to return the items for recycling. In conjunction, bottles and cans in certain states must clearly display a recycling logo representing the type of material used and providing easy-to-understand and transparent information about recycling. Finally, various regions have banned the inclusion of bottles and cans in solid waste – a command-and-control form of regulation.

Market Mechanisms

Market instruments have been used to address a variety of environmental issues in North America. The most recent have targeted air quality and climate change and include an acid rain reduction programme, a greenhouse gas emissions trading programme in the northeastern states and eastern provinces, and a carbon tax in Quebec and British Columbia. Payment for ecosystem services is also gaining wider attention, although such schemes remain limited.

Early projections of the average cost for the first phase of the programme ranged from a high of US$307 per tonne of sulphur dioxide removed to US$180 per tonne (1995 dollars). Ellerman *et al.* estimated that the actual costs were closer to the low end of the projections, in the range of US$186–210 per tonne. In addition, a 2011 US Environmental Protection Agency (EPA) review of the direct benefits to human health and the environment of the Clean Air Act estimates that these will reach almost US$2 trillion by 2020 while implementation costs are US$65 billion – a benefit-cost ratio of 30:1. This was probably due to the flexibility aforded to producers to find low-cost compliance measures, although other factors such as unanticipated technical improvements, lower transport costs and increases in coal production and use efciencies also played important roles. Although the costs of many regulatory programmes tend to be overestimated while they are being developed, recent research found that this has been especially the case for market-based programmes.

The success of the sulphur dioxide trading programme has in part prompted several jurisdictions in Canada to increase the use of market-based instruments. As of 2007, the Alberta greenhouse gas emissions trading system, for example, requires large industrial emitters that have been established more than eight years to reduce the intensity of greenhouse gas emissions by 12 per cent per year relative to a 2003–2005 baseline, and purchase carbon offsets or else pay a tax of US$15 per tonne of CO_2-equivalent. While the programme may result in reduced emissions compared to the business-as-usual alternative, it has been heavily criticised for permitting overall increases in carbon emissions by only targeting emissions intensity. In this sense it is not a typical cap-and-trade programme.

A less developed scheme, but one that is emblematic of the readiness of some states and provinces to compensate for perceived federal inaction, is the Western Climate Initiative, which combines seven US states and four Canadian provinces.

This has been working since 2007 to develop policies to address climate change, including a regional, economy-wide cap-and-trade programme and forest offset mechanisms. Only some of the initiative's members – California, Quebec and British Columbia – are currently taking preparatory steps towards implementing this programme in 2012.

Water trading between Canada and the United States and efforts to allocate water efciently and equitably among various users have triggered considerable political controversy, even before the United Nations acknowledged access to clean water and sanitation as a fundamental human right in 2010. Trading water rights, from farms to cities, for example, can be viewed as making farmland unproductive and favouring urban dwellers over rural residents. In addition, many civil society organizations see the privatization of some water rights as incompatible with the principle of universal and equal access to water.

Water markets, or transferrable water rights, are generally most developed in regions where water allocation is based on first-in-time, first-in-right or the doctrine of prior appropriation. In the United States, water markets are prevalent in the arid western states, and in Canada, water trading occurs in Alberta and to a lesser extent in British Columbia and the Territories. The benefits of water trading include the reallocation of water from lower- to higher-value economic uses or from areas where the marginal value is low to where it is high. For instance, where urban users pay much higher rates for water than do rural and agricultural users, trading makes both water buyers and sellers better of economically. There are numerous drawbacks, however. For example, the market value of water may not correspond to its *in situ* environmental value. Moreover, the impact on local water may be externalized to third parties, including changes to the local economy and environmental effes from reduced local water availability. Other drawbacks pertain to the very principle promoted by some groups that water should remain a public good and therefore should not be commoditized and traded for profit, the ability of private parties to monopolize the water resources market, and the distortion of the water trading market due to substantial water subsidies for the agricultural sector.

Subsidies and tarifs for clean energy, agricultural production and industrial goods can facilitate the adoption of new, less polluting technologies or projects that enhance energy conservation. Subsidies for installing water-efcient fxtures or the California subsidy programme on residential solar installation, which encourages distributed electrical generation as well as emission-free power production, are two such examples. The Ontario Feed-in Tarif programme, enabled by the 2009 Green Energy and Green Economy Act, ofers stable prices for energy provided by renewable sources and supports Ontario's objective to phase out coal-fred electricity generation by 2014. This programme has contributed to greater reliance on renewable energy sources in Ontario, such as wind power, which increased from 15 megawatts in 2003 to more than 1 100 megawatts in 2009.

While subsidies may help promote technological change, they have also been criticized for increasing the risk of pollution, encouraging overconsumption, and fostering the rapid depletion of natural resources. Agricultural subsidies have come under the greatest scrutiny not only because of their pervasive environmental effes on land use, but also for their negative impact on the agricultural sector and exports of developing countries. Both Canada and the United States also continue to provide large subsidies for the production of nonrenewable energy, often in the form of low tax rates for capital investment, despite the commitment to the contrary made by the G20 economies in 2009 in Pittsburgh. While some potentially environmentally harmful subsidies may have social or other worthwhile objectives, many may not be equitable, may no longer fulfl their original purpose, or may have unintended outcomes as a result of market distortions. There are many instances where subsidies have either directly or indirectly distorted the market or caused unintended consequences: for example, declining block rate structures for water use, where marginal costs decrease as a function of the total amount of water used, encourage overconsumption.

Payment for ecosystem services, which in one form or another has been used for years but has lately triggered considerable renewed interest, is designed to safeguard or increase the provision of an ecosystem service for which there is high demand but currently no market mechanism. The US Conservation Reserve Program, which provides continuous direct payments to farmers for withdrawing land from production and engaging in soil restoration, is a long-standing and successful example. The US Economic Research Service (ERS) conservatively estimates the programme's benefits to be US$1.3 billion per year, excluding carbon sequestration, ecosystem protection and other less easily quantified benefits. Other significant ecological benefits include the reversal of landscape fragmentation, maintenance of regional biodiversity, creation of wildlife habitat and favourable changes in regional carbon flux. The Environmental Quality Incentives Program and the Conservation Security Program of 2002 are two more recent and wide-ranging programmes that seek to reward farmers for sound land management from a multi-functionality perspective. For the same budgetary outlay, the ERS found that environmental performance could improve 12-fold, including an estimated 17 per cent reduction in soil erosion – saving about 36 million tonnes of soil valued at about US$2 per tonne, although the value of reducing sheet and rill erosion alone could be as high as US$332 million when in-stream sediment decreases are included. In addition, nitrogen leaching declined by 14 per cent, nitrogen run-of by 13 per cent, phosphorus run-of by 15 per cent, soil productivity losses by more than 300 per cent, wind erosion by 21 per cent, carbon emissions by 7 per cent, pesticide leaching by 9 per cent, and pesticide run-of by 7 per cent. The US Department of Agriculture has formed an Ofce of Environmental Markets to create guidelines for developing these kinds of market-based policies.

In Canada, continuous direct payment programmes based on a multi-functionality approach remain uncommon. Some provinces are already using payment for ecosystem services to make it more attractive for farmers to maintain stream habitats, while at the national level efforts are under way to find approaches for comparing the value of services provided by forests. The implementation of such schemes faces numerous methodological, political and ethical challenges as well as capacity, cost and time constraints, and their long-term impact is still unclear. In general, payment for ecosystem services needs to be complemented with land-use planning frameworks to be effective.

One innovative and promising economic approach aims to reduce the financial risk of switching to more environmentally sound practices and does not necessarily involve any payment. For instance, in the Canadian province of Prince Edward Island, farmers were ofered insurance against the perceived risk that reducing fertilizer use might also reduce yields. In the majority of cases, no payment was needed since reducing fertilizer use did not reduce yields: this was because fertilizer use was already so high that using less had little effe.

Command-and-Control Mechanisms

The use of public authority to preserve a given resource has a long and successful history. Changing private ownership to public or government ownership and a state-controlled protective regime can eliminate incentives to appropriate the benefits of overexploitation. Indeed, North America pioneered the establishment of the first national parks. This strategy presupposes extensive political and administrative enforcement of the status of these resources, which is more readily available in highly developed economies. Although its effectiveness remains to be seen, the Quebec Water Law of 2009, which considers water a common heritage of the Québécois nation, is a recent and noteworthy example of this type of instrument.

Command-and-control mechanisms are often preferred when there are significant threats to human health, when a specifc requirement needs to be monitored and enforced, when absolutely no additional environmental harm is permitted, and when simplicity and consistency are desired. In practice, market-based and command-and-control style regulations are often combined to meet an environmental objective. The ban on leaded petrol in the United States, for example, was accompanied by a trading mechanism during the phase-out period so that refneries could meet the declining production allowance in a cost-effective manner.

Although such instruments have become politically challenging to put together, particularly in the United States, there are several noteworthy examples of their successful use, such as standards for drinking water, clean air, toxic chemical releases and fuel; various types of prohibitions including on littering and the introduction of invasive alien species; and requirements on recycling, for example. Canada has the authority to regulate toxic substances, several fuels including diesel and petrol, and a number of fuel quality parameters, including sulphur levels. Greenhouse gas and air pollution regulations have also been implemented in Canada and the United States for new vehicles and engines. With regard to air quality control more generally, Canada monitors and regulates air pollutants through the Canadian Environmental Protection Act and has established National Ambient Air Quality Objectives, although air quality remains the primary responsibility of provinces. In the United States, the Corporate Average Fuel Economy (CAFE) standard regulates the fuel economy of new light-duty vehicles. as when penalties for failure to obey them are high enough, these changes usually depend on the continuous enforcement of regulations. Many governments at various levels have tried to green their operations, but the results have often been disappointing and have remained limited as long as they were perceived as top-down mandates and did not change the incentive structure. However, the positive experience of the US Forest Service since

2008, which sought to instil not only a conservation ethic but also a consumption ethic by changing organizational incentives and promoting bottom-up efforts, is instructive in this regard.

Accountability and Transparency

Policy instruments designed to increase accountability and transparency seek to make information on environmental performance and the environmental impacts of resource use more widely available to facilitate decision making as well as mobilize a variety of stakeholders. Certainly the best known and most widely disseminated of these policy tools is the requirement for environmental impact assessments, which, when first included in the 1969 US National Environmental Policy Act, mandated preliminary interdisciplinary assessments of the likely environmental impacts of major federal projects. It required US federal ofcials to include environmental values in a federal decision-making process dominated by technical and economic, if not political, considerations. An environmental impact assessment also requires the identification and evaluation of reasonable alternatives to a proposed federal action, as well as input from concerned stakeholders. Canada adopted its own act in 1992, following previous provincial initiatives. This has since evolved considerably, notably in terms of its target, which goes beyond federal and even publicly funded projects, but also in terms of its scope, with the introduction of sectoral and strategic assessments, and methods that include social variables. Although often criticized for its cost, the delays it can cause, and for ignoring the value of not doing anything at all (null decision), it remains one of the most effective tools for making sounder environmental decisions as well as improving participation.

The requirement to report on polluting emissions is another example of information dissemination that can become an effective policy tool. Canada has a National Pollutant Release Inventory and has implemented a Greenhouse Gas Emissions Reporting Program, while in the United States the EPA requires the reporting of greenhouse gas data and other relevant information from large sources and suppliers. This reporting is now required in 19 US states, and companies will need to report federally on their 2010 emissions in 2011. The US Toxics Release Inventory programme provides stakeholders with information about chemical releases for better decision making. The drawbacks of such instruments include the limited effectiveness of relying on blame-and-shame alone when inventory requirements are not tied to specifc obligations. Thus, this instrument is best seen as a complement to market-based or command-and-control approaches. Providing basic information on the environmental impact of individual citizen behaviour is another useful policy instrument.

The US EPA and Department of Energy instituted the Energy Star labelling programme to recognize appliances that perform at or above category benchmarks for energy efficiency. It confers a simple efficiency label to a product, but not detailed information about its energy consumption or the anticipated operating costs. The benefits include its simplicity, which led to the rapid improvement of product efciency by manufacturers who wanted to qualify for the EnergyStar label. In addition, the creation of third-party advisory bodies has proved useful in balancing the needs of science and politics, and provides a means of enhancing policy resilience, that is, the capacity of given policy objectives and means to persist in the face of external challenges. Nationally, the Committee on the Status of Endangered Wildlife in Canada (COSEWIC), where federal and provincial bodies work together with private ones to diagnose problems and recommend action, has protected wildlife conservation from the vagaries of political cycles.

North America has also pioneered the institutionalization of public participation, which helps increase the likelihood of a policy's implementation. Examples include the Great Lakes agreements, the Commission for Environmental Cooperation's process of citizen submission on enforcement matters, and environmental public hearings, as through Quebec's Environmental Public Hearing Ofce. Specifically, Articles 14 and 15 of the North American Agreement on Environmental Cooperation (NAAEC) provide a non-adversarial process that allows citizens to fle assertions that a Party of the NAAEC (Canada, Mexico or the United States) is failing to enforce its environmental law effectively. In some cases, this process can lead to a record. Political checks and declining funding for the Commission for Environmental Cooperation, which had held constant over the years, have however threatened its effectiveness.

A noteworthy development, reffeing a trend seen in other countries, has been the use of the Ofce of the Auditor General to evaluate and publicize the degree of implementation of national or sub-national commitments. Canada created a Commissioner of the Environment and Sustainable Development in 1995, and Quebec followed suit in 2006, with both enjoying a fair degree of autonomy. This role will be enhanced by the recent adoption of various sustainable development strategies, both at the federal and provincial/state levels, aiming to make environmental decision making more transparent and accountable. It is too early to assess effectiveness, however, and the lack of uniform sustainable development indicators hampers comparisons between the approaches adopted.

The acceptability, nature and effectiveness of various policy instruments depend on a number of internal and external factors that vary from state to state, province to province and region to region. In the end, successful policies rely on a mix of instruments and incentives. Although market approaches have raised considerable interest and have proved to be efcient in some cases, the enduring value of traditional command-and-control regulation, associated with disclosure requirements, has been most effective in changing the behaviour of major polluters.

Land Use and Management

One of the most important obstacles to sustainable land use in North America remains the fragmentary nature of land management. Forests, rangelands, croplands and urban, suburban and peri-urban lands are all part of the same landscape mosaic from which people derive survival and quality of life. Often, activities within one land type affect the state of others, as well as other ecosystem services such as air and water quality. Such impacts are often referred to as externalities, as the true costs and benefits of the impact are borne by parties external to the those who control and benefit from the activity. Even within a given land-use type, management responsibilities can be dispersed across several distinct bodies according to the type of activity taking place or to the component that is under consideration - water, fish and wildlife, fossil fuels or recreation, amongst others. In forest planning, for example, forestry, oil and gas, recreation, and the provision of ecosystem services are often managed by entirely separate bodies, even though all activities take place within the forest.

In North America, many land-use policies are gaining support and are now considered to be highly effective in motivating sustainable land use. These policy options work in tandem, providing informational and functional support to achieve the desired goals. These policy clusters are:

- implementing integrated land management plans to encourage and enable sustainable resource use;

- incorporating the true costs and benefits of ecosystem services when developing policy mechanisms; and
- improving planning for and sustainability of public lands.

Implementing Integrated Land Management Plans

To speed up achievement of the international goal of sustainably developing and using land in North America, integrated planning is crucial, requiring policies with clear agreed goals and specifc targets. Land-use policies need to be set at the appropriate geographic scale - state, province, county and city level - although watersheds or other ecologically relevant geographic scales may be the most logical units for determining a resource-use sustainability plan. Specifc targets should be set to obtain the highest benefits for the least social and economic costs. Institutional barriers such as centralized yet fragmented governmental structures should be overcome to allow a regional emphasis, and stakeholders should be allowed to participate in spatial planning. Both regulatory and incentive-based policies can be enacted to encourage target attainment. These policies should motivate individuals and corporations to act in accordance with the established plans. In addition, policies should be developed to encourage the resource sectors to maintain and enhance ecosystem resilience for future generations as well as limit the erosion of ecosystem services.

Jurisdictions throughout North America have adopted many of these policy instruments, to different degrees. For example, in British Columbia, resource companies, environmental groups and coastal First Nations have successfully carried out an ecosystem-based integrated land-use planning exercise, the 2006 Great Bear Forest Agreements, through a collaborative process, although the recent economic downturn has made fnancing participatory and multi-agency programmes more difficult for state/ provincial and local governments. As fscal issues may become more challenging in the near future, creative fnancing and regulatory measures coupled with financial incentives could become more important. At the same time, agencies will have more time to develop plans as the pace of industrial, commercial and housing development slows. Planning today may have long-reaching impacts as the economy returns to normal.

States, provinces, counties and cities have taken action to encourage smarter land use through innovative policy mechanisms. These initiatives address many of the challenges related to an optimal land-use pattern while respecting property rights, the need for equity and low-income housing, employment concerns, resource protection and environmental issues. For example, in the United States, the State of Maryland uses a series of incentives in its Smart Growth programme. The programme rewards people for relocating close to their place of employment, capitalizes on state money for infrastructure by providing it only within planned growth areas (priority funding areas), targets conservation funding to contiguous land and high-conservation-value land within clearly identified Rural Legacy areas, and subsidizes urban redevelopment through its brownfeld redevelopment programme. Smart Growth focuses on long-term regional considerations of sustainability, valuing community, public transport, employment and housing choices, preserving natural resources and promoting equity.

Similarly in Canada, the Province of Ontario has developed a greenbelt around the City of Toronto and protected open space and working lands from further conversion through zoning regulations. Agricultural retention can have economic, cultural and amenity benefits as well as environmental ones. British Columbia has designated an Agricultural Reserve, while Vancouver promotes development near its Sky Train stations. Rather than continue investments in roads and highways that promote an automobile culture, metropolitan areas like Toronto and Vancouver are focusing scarce investment on public transport and transit-oriented development with multiple benefits.

Ecosystem Services and Private Sector Decision Making

Market mechanisms, financial incentives and regulatory approaches have moved people to adopt better land-use practices. However, policies intended to benefit society can have unintended consequences. They often require the conversion of forests, grasslands and wetlands to other uses, which results in loss of habitat and biodiversity, impaired water quality, increased fooding, eroded soils and loss of resource-based industries and employment. Governments can help diminish such environmental effes through a number of policy initiatives. The most efcient and least controversial remains the establishment of mechanisms through which users of an ecosystem service, such as water quality, who are willing to pay for the service, compensate land managers for implementing best management practices such as riparian bufers, reduced tillage and reduced fertilizer applications. Taxes and other incentives in the United States have increased the total area conserved by local, state and national land trusts to almost 15 million hectares. Payment for ecosystem service programmes, such as working lands (agricultural and forest) preservation programmes that bring together the various economic and ecological benefits that these lands provide to society, have permanently preserved another 92 million hectares in the United States.

Cap-and-trade systems, such as the one in place for wetlands in the United States, can also be established when the users of the ecosystem services are dispersed or even do not yet exist, as in the case of acting in the interests of future generations. Caps need to be established, as in the case of the policy of no net loss of wetlands in the US Clean Water Act, and the magnitude and nature of compensation needs to be determined. While requiring considerable resources of time and effort to establish and implement, the pay-of from a societal point of view is that the market is then able to determine the most efcient means of respecting the cap through a system of trading. In the more than 500 wetlands mitigation banking schemes that generate US$3 billion dollars, and the more than 110 habitat banks generating US$370 million in the United States, land developers include the cost of wetland mitigation when pricing potential land acquisitions. They understand that purchasing land with wetlands will cost more in the end than land without them; either they protect the wetland or are required to restore wetlands elsewhere. Governments can implement programmes to encourage wetland restoration projects that developers can pay for and use to mitigate any wetlands destroyed in their own development projects.

Improving Sustainability on Public Lands

In both Canada and the United States, which are endowed with diverse and abundant land resources, the government owns a substantial amount of that land: 89 per cent of the land mass in Canada and 35–40 per cent of it in the United States. While human capital in both countries remains a tremendous asset, many economic sectors continue to generate wealth through natural resource use. Therefore, federal government policies on its own land can have a large impact.

In the United States, principles of multiple use and sustained yield dominated for many years, then in 1993 President Clinton established a goal of achieving sustainable forest management of all US forests by the year 2000. And in 1995, through the Montreal Process and the Santiago

Declaration, the United States committed to a process of developing and evaluating national indicators of sustainable forest management. As a result, during the passage of the Federal Ecosystem Management Initiative, its emphasis shifted to ecosystem management with plans focused on long-term sustainability rather than on management to maximize short-term yield. However, planning has proven problematic and litigious, and recently, a revised planning rule has been proposed for the nation's public lands. The latest planning rule under consideration stresses the restoration and maintenance of forests and grasslands; the protection of water quality and ecological integrity of riparian areas; habitat provision for plant and animal diversity and species conservation; multiple uses including recreation and industrial applications; public involvement in the planning process including community consultation and all levels of government entities; the use of the best available scientifc information to inform the planning process; and the development of a more efficient and adaptive land management planning process.

While the planning rule is being revised, some groups argue that, instead of a governmental planning approach to help the forest, some type of certification processes for land and management practices should be implemented. Examples of such processes include those used by non-governmental groups including the Forest Stewardship Council and the Marine Stewardship Council's fisheries certification programme. Indeed, the province of Quebec's revised Forest Act, which sets the stage for integrated land management with significantly increased responsibilities at the regional level, legislates for wood products from all public forests to be eco-certified by 2013.

Public-private partnerships have become increasingly important as current government funds and stafng are inadequate to assess resources, coordinate sustainable management, and accommodate the increasing demands of multiple users. Public-private partnerships are difficult to foster unless sufcient motivation exists on all sides, as within federal agencies and among their staf, long-term traditions can be difficult to alter without appropriate changes to incentives and reward structures.

Case Studies on Innovative Land-use Policies

The policies, underlying conditions and case studies presented here demonstrate that multiple policy instruments can speed up efforts to achieve the internationally agreed goal of implementing integrated land management and water-use plans to ensure the sustainable use of renewable resources. In the case of the State of Maryland, policies leveraged the state's funds to encourage built infrastructure in planned priority areas while providing incentives to create new jobs and develop brownfeld sites within the same areas. The planning process involved local communities and used incentives to encourage voluntary participation to achieve the plan's goals, ensuring that it was politically palatable and thus likely to be successful. While encouraging development in and near cities, Maryland also protected valuable resource-rich land from conversion through permanent conservation easements.

In the case of Ontario and British Columbia, their governments passed regulatory measures to protect environmentally sensitive and working lands while encouraging transit-oriented development within the cities. From a policy perspective, environmentally sensitive and working lands are lumped together, and farming and environmental communities have joined forces on these issues – one of the reasons for so much support for conservation programmes. Conservation practices can be adopted to retain topsoil and prevent erosion

Integrated Watershed Management

Integrated watershed planning and management can be applied in combination with other water management measures and has become an indispensible instrument for improving water resources. It is an holistic approach to managing water within drainage areas. Integrated watershed planning and management has proved effective in addressing some complex challenges over the last few decades. The method recognizes that water issues cannot be addressed independently but require the balanced consideration of all environmental, social, economic and technical aspects. It may include goals such as food prevention, enhancement of aquatic habitat and biodiversity, reduction in the loss and degradation of wetlands, pollution control and economic growth. The success of programmes can be assessed through water quality indicators including contaminant concentration, dissolved oxygen and biodiversity, water fow and food prevention.

Developing and implementing an integrated watershed planning and management policy requires active participation, interaction and collaboration between stakeholders. Currently, this is not administered nationally in the United States and Canada, but through initiatives at the regional or state/provincial level. For example, the Total Maximum Daily Loads programme for pollutant control in the United States is being implemented at the state level as required by the Clean Water Act. States are required to identify impaired waters and calculate the maximum amount of a pollutant a water body can receive and still meet water quality standards, and then develop plans, with public input, to address point and non-point sources of pollutants in an effort to restore and maintain the water quality. Although the programme has shown a varying degree of success across the country – due in part to the diferences of each watershed – factors that have been recognized to enhance implementation include a focused watershed plan, active stakeholder involvement, coordination between local and state governments, a diversity of approaches to addressing sources of pollution, and adequate resources for watershed characterization and monitoring. An attractive aspect of integrated watershed planning and management is that it need not require expensive infrastructure such as water treatment and control structures. Therefore, costs do not necessarily restrict implementation, so it can move forward in situations and regions where financial resources are limited. This makes integrated watershed planning and management highly transferable, provided effective coordination and implementation mechanisms can be established. It can also be applied at a diversity of scales ranging from small urban stream restoration projects to large watershed programmes, such as the Great Lakes, Chesapeake Bay, the Everglades and San Francisco Bay. Of its many benefits, perhaps the most notable is that stakeholders are actively involved in selecting the management strategies to solve water resource problems. Active stakeholder involvement, with explicit discussion of issues, improves decision making and acceptance, thus ofering advantages over top-down planning, which often lacks public support and understanding.

Integrated watershed planning and management is not without problems, however, and it is often difficult to determine how well it works. In the Chesapeake Bay watershed, it was initiated decades ago in an effort to clean up the estuary and restore coastal fisheries. Projects to improve water quality have largely focused on tributaries, and include re-vegetating riparian areas, improving stream channels and restoring wetlands. Millions of dollars have been spent on thousands of restoration projects within the watershed, yet the success is difficult to gauge, due in part to a lack of comprehensive monitoring of individual projects. While clear indications of widespread water quality improvements in the Chesapeake Bay have not yet been observed, outcomes in some areas look promising.

In general, integrated watershed planning and management faces serious challenges due largely to the magnitude and complexity of problems as well as socio-political rather than technological or hydrological barriers. The mismatch between watershed boundaries and political boundaries poses a challenge because of the often conficting needs of the multiple landowners and political entities with jurisdictions in watersheds. To overcome this, a watershed authority is typically established to coordinate and implement the plan, and faces the formidable task of bringing together the stakeholders and facilitating agreements to balance the needs of competing interests. Thus, collaboration and public participation are essential. The challenges of creating watershed authorities are magnified when watersheds cross international boundaries. However, these challenges can be met through such efforts as the International Watersheds Initiative, which was conceived by the governments of Canada and the United States to promote the establishment of watershed authorities and facilitate integrated transboundary management.

Full-cost Pricing

Full-cost pricing of water delivery has been defned by the US EPA as "a pricing structure which fully recovers the cost of providing that service in an economically efcient, environmentally sound, and socially acceptable manner, and which promotes efcient water use by customers". Based on the user-pays and polluter-pays principles, high-volume users pay proportionately more than low-volume users. The aim is to make it possible for all consumers to aford the volume of water necessary for basic human needs while charging increasing prices for consumption beyond that level. Full costs include all public and private costs, both market and non-market values, and account for costs that will be incurred in the future, such as those arising from infrastructure rehabilitation and replacement. In public water provision, once the water delivery infrastructure is in place - such as dams, canals, pumps, pipelines or treatment plants - the marginal cost to the utility company of delivering water to its customers is equal to its variable costs. These costs amount primarily to administrative and maintenance costs, which are near zero compared to the cost of establishing the overall infrastructure.

The resulting artifcially low market price to customers generally leads to water consumption decisions being based on incomplete information, resulting in overconsumption. In a full-cost pricing model, all infrastructural, environmental and intergenerational costs are included in the delivery price. In practice it is difficult to account for all of these costs accurately; nonetheless, various pricing systems attempt to convey more complete cost information so as to require consumers to pay more of the costs associated with their respective levels of water consumption. One example of how full-cost pricing can be implemented is through increasing block rates, thought to be the most effective in encouraging conservation. In this pricing structure, the amount charged per unit of water consumed increases with the total volume consumed.

Numerous examples of successful implementation of full-cost pricing exist, and are typically evaluated in terms of reductions in water consumption. An example is ofered by the Marin Municipal Water District (MMWD), a public agency that provides water for 195 000 residents in south and central Marin County, California. The MMWD's water rate structure includes a base fee that covers such services as meter reading, billing, meter replacement and repair, customer service, water conservation and administration, and four levels of charge that cover the cost of water transmission, treatment, distribution, watershed maintenance, and importing and recycling water. The MMWD imports a quarter of its water from the Russian River in Sonoma County through an agreement with the Sonoma County Water Agency. The environmental costs of using water from the Russian River stem from the Federal Endangered Species Act, and include expenses related to improving conditions for several fish species that are classified as threatened or endangered, for example by constructing fish ladders, as well as channel maintenance and monitoring. The MMWD is unusual in that customers pay the full cost of water without state and federal subsidies or cost sharing with other water agencies. Rates are comparable to other northern California water agencies, and overall water use has remained relatively stable over the last several decades despite an increasing population. These water-saving measures are a result of a better understanding of the true value of water, and have minimized the financial and environmental costs of water supply expansion.

Despite successes, there are also some limitations to full-cost pricing, including its complexity compared to the simplicity of traditional marginal-cost pricing structures, making it difficult for consumers to respond to the price information by adjusting their water use. Public outreach campaigns and in-bill information leafets that describe cost structures are addressing this barrier to some extent. Another limitation is the difficultly in setting prices properly, in particular in identifying and allocating non-market costs such as environmental losses associated with water delivery – for example the environmental effes of constructing new diversion and containment structures. Various formal methods have been developed, however, for assigning market values to non-market costs over time, identifying present and amortized values for those costs, and then adding them to the marginal cost to customers on the basis of water consumed. Implementing full-cost pricing requires adequate institutional support and agreement, as well as the personnel and data necessary for estimating cost components.

Technological Solutions and Conservation Measures

Technological advances and conservation measures can effectively decrease water use in the residential, industrial and agricultural sectors. This has been accomplished in large part through regulation, financial incentives and voluntary measures. Many options are available for reducing water consumption and increasing efciency depending on the sector, including low-tech solutions, water-saving appliances, water reuse systems and metering. For example, the decline in average residential water use in North America over the last 25 years is largely attributed to increased efciency standards for household appliances. In the agricultural sector, food irrigation systems are being replaced by more efcient technologies designed to increase crop yield per unit of water use. Simpler conservation measures such as responsible water-use habits go hand-in-hand with efciency, and can be promoted through water education programmes. Examples of cities that have implemented such programmes include El Paso, Texas, San Diego, California and Prince George, British Columbia.

Conserving water through improved long-term sustainable efciency can lead to a range of economic and environmental benefits. Some of the advantages of this approach include adaptability to site-specifc needs, avoidance of more expensive potable water supplies, and the reduced costs of operating and maintaining water distribution and treatment infrastructure, with associated energy savings. For commercial and industrial facilities, savings in water and energy costs realized by implementing efciency measures can quickly offset the investments made. For instance, in the State of California the average estimated payback period for investing in water-efcient technologies in the commercial, industrial and institutional sector is typically less than two and a half years. Obstacles to implementing water efciency measures include situations in which the capital investment does not justify the water cost savings in the short term, or when a general consensus cannot be reached among stakeholders that the benefits accrued to the water rate payers are worth the investment in the long run. Decisions often depend on the costs associated with water use and water discharge, environmental compliance and production. Other economic incentives may be required in some areas with low water costs, including subsidies, tax credits and grants. In many cases, it will be a combination of sector-specifc instruments and incentives appropriate to a region's issues and needs that will allow a variety of innovative and effective water-use efciency measures to be implemented.

Policies that promote the integrity of the water cycle and the essential life-supporting services it provides can indirectly help achieve the internationally agreed goals for land use and renewable energy. Successful implementation of integrated watershed planning and management is likely to promote sustainable land use by restoring ecosystem function and enhancing resilience. When the true cost of water supply is assessed, added revenue may be used to fund restoration programmes carried out over the landscape. Water conservation that stems from financial incentives and technological advances will further reduce land degradation and minimize energy requirements for the use and distribution of water. Greater reliance on renewable energy sources will reduce greenhouse gas emissions that cause climate change, which may mitigate projected impacts on the water cycle.

Management of Energy Crisis

Canada and the United States are endowed with diverse and abundant renewable energy resources. Transforming that vast potential into a sustainable energy system requires mobilizing political will, behavioural change and smart, comprehensive policies that support renewable energy. There are several environmental issues associated with the current energy system, including climate change, elevated water consumption and air pollution. Since fossil fuel consumption is the major contributor to increasing atmospheric concentrations of carbon dioxide (CO_2), experts contend that policy interventions should be strengthened, not just to increase renewable energy production, but to substitute renewable energy for the current carbon-emitting energy systems. Renewable electricity technologies ofer an effective means of reducing greenhouse gas emissions, thus providing a tool for climate change mitigation. It has become clear that even partial mitigation of the rate of climate change requires more carbon-free sources of electricity. In addition, policy innovation and technical improvements are rapidly advancing in this sector, thus providing the clearest examples for emulation.

North America's current dependence on fossil fuel resources largely stems from a cycle of pricing effes, partially due to subsidies that favour conventional fossil fuel energy production and that externalize pollution costs. For example, an analysis of all energy subsidies provided in the United States in 2004 shows that 86 per cent went to fossil fuels, 8 per cent to nuclear energy and just 6 per cent to renewables and energy efciency. Recently, Energy Secretary Steven Chu announced that the Obama administration intends to repeal US$46.2 billion in subsidies to oil, natural gas and coal companies in the next ten years in order to fund renewable energy spending. Economists argue that to address these uneven subsidies and other market failures associated with fossil fuels and to accelerate renewable energy deployment, the multiple social and environmental costs of emissions have to be included in the price of conventional energy production. Smart, novel and comprehensive policies are therefore necessary to provide the incentives, transmission networks, transparency and market space essential to support rapid and sustained renewable energy development and the substitution of fossil fuels.

During the selection process, three policy clusters were identified affeing renewable energy adoption: providing financial support to alter incentives or encourage behavioural change; improving networks and grid flexibility; and decreasing institutional barriers. Such an approach could accelerate renewable energy development by simultaneously confronting the multiple challenges and barriers that are delaying the transition to a sustainable energy system.

Support to Alter Incentives or Encourage Behavioural Change

Examples already in use in North America include production tax credits, feed-in tarifs and renewable portfolio standards; in addition, governments supply funding for research and development. Production tax credits represent kilowatt-hour tax credits for qualified renewable energy sources while feed-in tarifs typically guarantee grid access and provide long-term contracts for electricity generation at stable prices. Where they are well designed, feed-in tarifs also provide renewable energy premiums using the rate-payer base rather than government funds. Renewable portfolio standard policies also avoid the use of government funds, with the exception of monitoring compliance with the standard, and typically require utilities to procure renewable energy resources as a prescribed percentage of total electricity. Investments in research and development help to improve technologies that drive prices down, providing market advantages aimed at increasing the renewable energy market. The close coupling of research and development with investment subsidies has shown to improve policy effectiveness.

Improving Networks and Grid Flexibility

Renewable energy sources and current fossil fuel generation facilities are often located in different places, thus requiring networks to transport energy from new source areas to load centres. In addition, fossil fuel generation, which is characterized by long-term capital stock, currently dominates the market, limiting opportunities for new technologies to enter. Several policy measures have been devised that improve the management and characteristics of transmission networks and increase market access and space. These include designating transmission cost recovery and allocation; managing the grid through independent system operators; developing smart grids; and phasing out coal plants. These policies are intended to make it easier to develop infrastructure, open market space and transmit renewable energy from areas of generation to load centres.

Cost recovery and allocation policies provide clear frameworks for developers to recover installation costs from transmission projects, which is necessary to provide an energy transportation network to increase renewable energy use. Currently, it is difficult to fnance the development of transmission structures that cross multiple state and provincial jurisdictions, with, in many cases, associated problems in assigning costs and benefit levels. To overcome this, experts have proposed that federal authorities should determine cost allocation.

Energy developers also encounter problems with the lack of transparency and access to the grid as, traditionally, vertically integrated companies generate, transmit and distribute electricity. In many areas, utility companies still own and operate the transmission assets, leading to a lack of transparency in the availability of transmission. Independent system operators are third-party public institutions responsible for granting access to transmission grids, which could provide desirable conditions for accelerating renewable energy deployment by ensuring transparency and fair access to markets. In Texas, where cost allocations are assigned to all supply entities, representing a novel approach for North America, the construction of high-voltage electricity transmission is proceeding rapidly.

Phasing out coal plants is a relatively new policy instrument that decreases greenhouse gas emissions while simultaneously increasing grid flexibility and providing market space for renewable energy. Since coal-fred technology has a limited ability to respond to load fuctuations, these policies typically substitute coal-fred generation with natural gas, which has more responsive technologies that emit lower levels of pollutants and greenhouse gases than coal-fred generation. Coal phase-out policies provide public health benefits and accelerate the transition to a sustainable energy system by decreasing emissions that lead to climate change. This particular policy rapidly internalizes the costs associated with the market failure of fossil fuel energy by targeting concentrated sources of emissions.

Policies for Overcoming Institutional Barriers

The final cluster consists of policies that increase the pace of renewable energy deployment by removing institutional barriers and facilitating long-term planning. One method of removing barriers is by consolidating siting authorities, either by aggregating multiple jurisdictions into one decision-making body or by placing the siting authority in an existing entity; examples are the Province of Ontario and the State of Texas.

Agencies may also conduct integrated resource planning, which typically requires involving the public, identifying energy efciency and resource options, developing action plans, and describing efforts to minimize the environmental effes of resource acquisitions. Experts contend that plans for designing and optimizing systems should now include explicit consideration of grid-connected renewables. They also maintain that including the evaluation of renewable energy sources in integrated resource planning helps develop a cost-effective sustainable energy system.

Benefits of The Selected Policy Measures

Empirical evidence shows that widespread renewable energy results in decreased environmental impacts and increased social benefits. Thus, increasing renewable energy production and displacing fossil fuels in the energy system by addressing perverse subsidies, providing paths to markets and market space, and removing institutional barriers could deliver multiple benefits. Environmental benefits include reduced greenhouse gas emissions and air pollutants, lower water use in the case of wind and solar photovoltaics, and decreased water pollution. Social benefits include enhanced energy security and reliability by diversifying the supply and using indigenous resources, and reduced energy price volatility and disruptions. In addition, experts maintain that renewable energy developments are associated with enhanced economic development and more jobs. Finally, the use of renewable resources also benefits public health through decreased emissions and fewer occupational injuries.

Research clearly demonstrates that renewable energy sources generate significantly lower greenhouse gas emissions than fossil fuel options. Scenario analyses indicate that increasing renewable energy deployment from 27 to 77 per cent of the primary energy supply by 2050 may be expected and may achieve savings of up to 85 per cent of global CO_2 emissions for the scenarios with the highest renewable energy shares. The majority of the technologies deployed in these scenarios are wind, direct solar and modern biomass, with an annual average cost of less than 1 per cent of global gross domestic product (GDP) per year. Furthermore, experts forecast that by 2030 the production costs, including social costs, of renewable energy would be lower than energy production by fossil fuels. However, to achieve this transition, existing policies must be significantly strengthened and implemented comprehensively and therefore require additional political will.

The benefits of improving networks and reducing institutional barriers include lower costs and faster deployment of renewable energy. In the case of transmission, improved networks generally enhance reliability, lower the delivered costs of electricity and restrict the ability of generators to exercise market power. Experts commonly call for reducing institutional barriers to expedite the transition to a sustainable energy system. Quantitative analysis also shows that reducing siting barriers correlates with increased wind power development.

Potential Drawbacks of Selected Policy Measures

The successful implementation of production tax credits or feed-in tarifs requires an in-depth understanding of the various energy prices for all renewable energy sources as well as the costs of externalities. These policies therefore have potential drawbacks. Specifically, production tax credits or feed-in tarifs can be extremely inefcient. Since incentive levels are fxed over time, this may lead to limited innovation and downward price pressures. Likewise, implementing renewable portfolio standards also requires an in-depth knowledge of markets to establish appropriate targets, enforcement mechanisms and sector-specifc set-asides. While context dependent, these are usually subsidies aimed at a particular industry. Inadequately designed renewable portfolio standards may encourage particular technologies and therefore lead to technological lock-ins.

In addition, critics argue that implementing renewable energy policies may increase the cost of energy and/or increase tax burdens. These expenses are especially burdensome to lower-income households; however, widespread renewable energy adoption combined with progressive tax design and incentives ofers some protection from energy price increases. For example, subsidy programmes already exist to assist low-income households with energy costs in the United States, so expanding existing programmes could provide assistance for vulnerable groups should energy prices rise.

Policies to increase transmission networks and reduce siting barriers also have potential drawbacks. When reallocating the costs of transmission, these policies could result in disproportionate financial burdens on parties who do not benefit. Reducing siting barriers may also decrease public participation.

Replication and Transferability of Selected Policies

The potential for replication and transferability of these policies is not straightforward and is arguably dependent on context and specifc instrument design. For example, the North American grid exists in an institutional framework that is highly fragmented, while other countries may have nationally owned networks, in which case fragmentation may not be an issue. Germany, France, Italy, Japan and Denmark have experience in replicating and transfering feed-in tarifs at the national level, while the United States and Australia have experience with production tax credits and renewable portfolio standards. Policies on feed-in tarifs and renewable portfolio standards are in force in diverse jurisdictions including Canada, China, Kenya, Portugal and Uganda. Statistically, correlations demonstrate that the policies are effective, particularly in the case of feed-in tarifs. Direct causal evidence of effectiveness for other policies, however, is limited, as is evidence of the potential for replication and transferability to other jurisdictions.

Proactive Measures to Accelerate the Use of Renewable Energy

Achieving the international goal of urgently expanding the share of renewable energy supply in North America's energy mix requires mobilizing political will and increasing public support to implement comprehensive renewable energy policies focused on addressing market failures, providing clear market signals, modernizing transmission systems, proving new technologies including energy storage, and streamlining institutional structures. A modernized, clean, reliable and efcient 21st century energy system will provide greater energy security, enhanced price stability and increased economic performance, and may save up to 85 per cent of global greenhouse gas emissions by 2050.

Current research argues that in accounting for fossil fuel externalities and subsidies, the question appears not to be one of cost, but rather of social and political barriers. Cultivating and developing widespread public participation and support is essential for generating the political will to implement the policies necessary to achieve the internationally agreed goal. The case studies illustrate that comprehensive policy packages that include incentives to offset externality and subsidy advantages aforded to fossil fuels, provide for energy transmission and reduce institutional barriers, can also significantly accelerate the transition to a sustainable energy future.

Increasing the deployment of renewable energy can provide a number of benefits to support the other internationally agreed goals. Wind and solar photovoltaic renewable energy can decrease water stress since it uses less water than conventional thermo-electric forms of generation. Benefits for land use include relative reductions in greenhouse gas emissions, thereby decreasing potential climate change impacts. However, land use for expanding renewable energy systems may require the disturbance of additional areas, depending on the particular technology being deployed. At the same time, an integrated approach to siting renewable energy, increased transparency and collaboration between agencies may lead to improvements in environmental governance.

References

Awerbuch, S. (2006). Portfolio-based electricity generation planning: policy implications for renewables and energy security. *Mitigation and Adaptation Strategies for Global Change* 11,693710

Carley, S. (2009). State renewable energy electricity policies: an empirical evaluation of effectiveness. *Energy Policy* 37, 30713081

Cortner, H. and Moote, M. (1999). *The Politics of Ecosystem Management.* Island Press, Washington, DC

Haas, R., Resch, G., Panzer, C., Busch, S., Ragwitz, M. and Held, A. (2011). Efficiency and effectiveness of promotion systems for electricity generation from renewable energy sources: lessons from EU countries. *Energy* 36, 21862193

Ritter, W.F. and Shirmohammadi, A. (2001). *Agricultural Non-Point Source Pollution: Watershed Management and Hydrology.* Lewis Publishers, New York

Sovacool, B.K. and Watts, C. (2009). Going completely renewable: is it possible (l*et al*one desirable)? *The Electricity Journal* 22, 95111

Spieles, D.J. (2005). Vegetation development in created, restored, and enhanced mitigation wetland banks of the United States. *Wetlands* 25, 5163

Bibliography

Abdulla, A., Gomei, M., Maison, E. and Piante, C. (2008). *Status of Marine Protected Areas in the Mediterranean Sea*. IUCN, Malaga and WWF, France

ACB (2011). *Legal and Institutional Development for Promoting Access and Benefit Sharing of Genetic Resources in Southeast Asian Countries*. ASEAN Centre for Biodiversity, Los Banos

ACIA (2005). *Arctic Climate Impact Assessment.* Cambridge University Press, Cambridge.

Ali, M.H. (2010). *Fundamentals of Irrigation and On-Farm Water Management* Volume 1, and *Practices of Irrigation and On-Farm Water Management* Volume 2. Springer Science+Business Media, New York, NY.

Andrade Perez, A. (ed.) (2008). *Applying the Ecosystem Approach in Latin America.* (translator Medina, M.E.). IUCN, Gland

Ashmore, M.R. (2005). Assessing the future global impact of ozone on vegetation. *Plant, Cell and Environment* 28, 949964.

Awerbuch, S. (2006). Portfolio-based electricity generation planning: policy implications for renewables and energy security. *Mitigation and Adaptation Strategies for Global Change* 11,693710

Bakker, M.M., Govers, G., Kosmas, C., Vanacker, V., van Oost, K. and Rounsevell, M. (2005). Soil erosion as a driver of land-use change. *Agriculture, Ecosystems and Environment* 105(3), 467481.

Barnett, J. and Adger, W.N. (2007). Climate change, human security and violent conflict. *Political Geography* 26, 639655.

Basel Convention (1989). *The Basel Convention on the Control of Transboundary Movement of Hazardous Wastes and their Disposal.* http://www.basel.int/

Bates, B., Kundzewica, Z.W., Wu, S. and Palutikof, J. (eds.) (2008). *Climate Change and Water.*

Bennett, A.F. (2003). *Linkages in the Landscape: The Role of Corridors and Connectivity in Wildlife Conservation.* Second edition. IUCN, Gland, Switzerland and Cambridge

Bongaarts, J. (2001). *Household Size and Composition in the Developing World.* Population Council, New York

Carley, S. (2009). State renewable energy electricity policies: an empirical evaluation of effectiveness. *Energy Policy* 37, 30713081

Carr, D. (2009). Population and deforestation: why rural migration matters. *Progress in Human Geography* 33(3), 355378

CBD (2004). *The Ecosystem Approach (CBD Guidelines).* Convention on Biological Diversity, Montreal

COE (2000). *European Landscape Convention.* European Treaty Series No.176. Council of Europe,Strasbourg

Corbera, E., Brown, K. and Adger, W.N. (2007). The equity and legitimacy of markets for ecosystem services. *Development and Change* 38(4), 587613.

Cortner, H. and Moote, M. (1999). *The Politics of Ecosystem Management.* Island Press, Washington, DC

Dietz, T., Fitzgerald, A. and Shwom, R. (2005). Environmental values. *Annual Review of Environment and Resources* 30, 335372

EEA (2011a). *Greenhouse Gas Emissions in Europe: A Retrospective Trend Analysis for the Period 1990–2008.* EEA Report No 6/2011. European Environment Agency, Copenhagen

Emerson, C. (1999). *Aquaculture Impacts on the Environment.* Cambridge Scientific Abstracts.

Enerdata (2011). *Global Energy Statistical Yearbook.* Enerdata, Grenoble

Eyring, V., Shepherd, T.G. and Waugh, D.W. (2010). *SPARC Report on Evaluation of Chemistry-Climate Models.* SPARC Report No. 5. Stratospheric Processes And Their Role In Climate. WCRP-132, WMO/TD-No. 1526.

FAO (2010). *Global Forest Resources Assessment.* Food and Agriculture Organization of the United Nations, Rome

Ferraro, P. (2001). Global habitat protection: limitations of development interventions and a role for conservation performance payments. *Conservation Biology* 15, 9901000

Finlayson, C.M. and DfCruz, R. (2005). Inland water systems. In *Ecosystems and Human Wellbeing: Current State and Trends: Findings of the Condition and Trends Working Group* (eds.Hassan, R., Scholes, R. and Ash, N.). pp.551583. Island Press, Washington, DC

Forero, E.G. (2008). The EA and water management: a Latin American perspective. In *Applying the Ecosystem Approach in Latin America* (ed. Andrade Perez, A.) (translator Medina, M.E.). IUCN, Gland

Frantzeskaki, N. and Loorbach, D. (2010). Towards governing infrasystem transitions: reinforcing lock-in or facilitating change? *Technological Forecasting and Social Change* 77, 1292130.

Glennie, P., Lloyd, G.J. and Larsen, H. (2010). *The Water-Energy Nexus: The Water Demands of Renewable and Non-Renewable Electricity Sources.* DHI, Horsholm.

Haas, R., Resch, G., Panzer, C., Busch, S., Ragwitz, M. and Held, A. (2011). Efficiency and effectiveness of promotion systems for electricity generation from renewable energy sources: lessons from EU countries. *Energy* 36, 21862193

Hansen, J., Ruedy, R., Sato, M. and Lo, K. (2010). Global surface temperature change. *Reviews of Geophysics.* 48, RG4004

Hoekstra, A.Y. and Mekonnen, M.M. (2011). *Global Water Scarcity: Monthly Blue Water Footprint Compared to Blue Water Availability for the World's Major River Basins*. Value of Water Research Report Series No.53. UNESCO-IHE, Delft.

IAEA (2009a). *Classification of Radioactive Waste General Safety Guide*. Series No. GSG-1. International Atomic Energy Agency, Vienna

IGES (2008). *Climate Change Policies in Asia-Pacific: Re-Uniting Climate Change and Sustainable Development.* White Paper. Institute for Global Environmental Strategies,

IPCC Technical Paper VI. IPCC Secretariat, Intergovernmental Panel on Climate Change, Geneva.

Kanji, N., Toulmin, C., Mitlin, D., Cotula, L., Taoli, C. and Hesse, C. (2006). *Innovation in Securing Land Rights in Africa: Lessons from Experience*. International Institute for Environment and Development, London

Kumar, P. (ed.) 2010. *The Economics of Ecosystems and Biodiversity: Ecological and Economic Foundations*. Earthscan, Washington.

Levin, S.A. (1998). Ecosystems and the biosphere as complex adaptive systems. *Ecosystems* 1,431436.

Levy, M.A., Haas, P.M. and Keohane, R.O. (1993). Improving the effectiveness of international environmental institutions. In *Institutions for the Earth: Sources of Effective International Environmental Protection*. MIT Press, Cambridge, MA.

MA (2005). *Ecosystems and Human Well-being: Synthesis.* Millennium Ecosystem Assessment.Island Press, Washington, DC.

Maretti, C.C. (2003). *Protected Areas and Indigenous and Local Communities in Brazil*. WCPA Ecosystems, Protected Areas and People (EPP) project. IUCN, Gland

May, P. and Millikan, B. (2010). *The Context of REDD+ in Brazil: Drivers, Agents and Institutions*. Center for International Forestry Research (CIFOR), Bogor

Medina, M. (2007). *The World's Scavengers: Salvaging for Sustainable Consumption and Production*. Alta Mira Press, Lanham, MD

Molden, D. (ed). (2007). *Water For Food, Water For Life: A Comprehensive Assessment of Water Management in Agriculture*. Earthscan, London and Water Management Institute, Colombo.

Montgomery, D.R. (2007). Soil erosion and agricultural sustainability. *Proceedings of the National Academy of Sciences of the United States of America* 104(33), 1326813272.

Nelson, F. (2010). *Community Rights, Conservation and Contested Land. The Politics of Natural Resource Governance in Africa*. Earthscan, London

OECD (2006). *Good Practices in the National Sustainable Development Strategies of OECD Countries.* Organisation for Economic Co-operation and Development, Paris.

Pereira, L.A.S., Cordery, I. and Iacovides, I. (2009). *Coping with Water Scarcity: Addressing the Challenges.* Springer Science.

Perret, S., Stefano, F. and Rashid, H. (eds.) (2006). *Water Governance for Sustainable Development: Approaches and Lessons from Developing and Transitional Countries.* Earthscan, London.

Peters, G.P. and Hertwich, E.G. (2006). The importance of import for household environmental impacts. *Journal of Industrial Ecology* 10(3), 89110

Ragwitz, M., Winkler J., Klessmann, C., Gephart, M. and Resch, G. (2012). *Recent Developments of Feed-in Systems in the EU – A Research Paper for the International Feed-In Cooperation.* A report commissioned by the Ministry for the Environment, Nature Conservation and Nuclear Safety (BMU), Bonn

Ritter, W.F. and Shirmohammadi, A. (2001). *Agricultural Non-Point Source Pollution: Watershed Management and Hydrology.* Lewis Publishers, New York

Seto, K.C., Sanchez-Rodriguez, R. and Fragkias, M. (2010). The new geography of contemporary urbanization and the environment. *Annual Review of Environment and Resources* 35, 167194

Shekdar, A. (2009). Sustainable solid waste management: an integrated approach for Asian countries. *Waste Management* 29(4), 14381448

Sherman, K. and Hempel, G. (2008). *The UNEP Large Marine Ecosystem Report: A Perspective on Changing Conditions in LMEs of the World's Regional Seas.* United Nations Environment Programme, Nairobi.

Sovacool, B.K. and Watts, C. (2009). Going completely renewable: is it possible (*let alone* desirable)? *The Electricity Journal* 22, 95111

Spieles, D.J. (2005). Vegetation development in created, restored, and enhanced mitigation wetland banks of the United States. *Wetlands* 25, 5163

Stalk, A. (2004). *Management of the Free Basic Water Policy in South Africa.* Master project. Roskilde University, Roskilde

Stern, N. (2007). *The Economics of Climate Change: The Stern Review.* Cambridge University Press, Cambridge and New York.

UN (2000). *Millennium Development Goals.* United Nations http://www.un.org/millenniumgoals/

UNCED (1992a). *Rio Declaration on Environment and Development.* United Nations Convention on Environment and Development, Rio de Janeiro

World Bank (2006). *Where is the Wealth of Nations? Measuring Capital for the 21st Century.* World Bank, Washington, DC

Ziolkowska, J. (2009). Environmental benefit, side effects and objective-oriented financing of agri-environmental measures: case study of Poland. *International Journal of Economic Sciences and Applied Research* 2, 7188